This Booklet is to
of a cruise on the
on the Thamesgr

JUNE 1992

A VICTORIAN VISION

The Ipswich Wet Dock Story

Overleaf: *John Moore's oil painting The Riverfront at Ipswich Docks captures the character of the dock area in the 19th century.* Ipswich Borough Museums and Galleries

A VICTORIAN VISION

The Ipswich Wet Dock Story

Bob Malster and
Bob Jones

IPSWICH PORT AUTHORITY, 1992

Published by
Ipswich Port Authority
ISBN 0 901029 09 2

British Library Cataloguing-in-Publication Data

A catalogue record for this book is
available from the British Library

Designed by Robert Malster
Printed in Great Britain by
Ancient House Press, Ipswich, Suffolk

Contents

Foreword

A HUNDRED and fifty years ago, at the very start of Queen Victoria's long reign, the Ipswich Dock Commissioners had the foresight and the vision to construct what was then the largest Wet Dock in the country—larger than any dock in London or Liverpool.

At a cost originally quoted at £65,178—a huge sum in those days—the River Orwell was redirected through the New Cut, an island of 14 acres was formed and the 33-acre Wet Dock was created.

What prompted the Commissioners to take such a brave decision, how was the dock built, and how did this venture fare? We hope this book will give the answers to these and other fascinating questions, and, by recalling the events during these years of Victorian enterprise, will add something to the historical records of the town of Ipswich.

In the 150 years since then, the Port of Ipswich has thrived despite some difficult periods; and, firmly set on the foundation of the original dock, has developed into one of Britain's leading ports. Over the years, other brave decisions have been made to extend the Port well beyond the dock along both banks of the river. The Port has made a major contribution to the prosperity of the region and the town.

While the 150th Anniversary of the Wet Dock affords us the opportunity to look back with pride, we believe that the Port can also look forward to the future with confidence and a determination to continue the good work so ably begun by our Victorian forebears.

John Austin
Chairman

Introduction

IPSWICH has been a port for over a thousand years. Its history is rich and fascinating. Many books have been written tracing the town's origins to Anglo-Saxon times and describing in detail the fortunes of the town and the trade of the port from those years onward.

This book commemorates one of the most significant events in the proud history of Ipswich as a port: the opening of the Wet Dock just 150 years ago. When the first ship passed through the lock into the new dock in January 1842, the 33-acre dock was the largest area of water of its kind in England. It helped to promote the growth of the town's industries during a period of great expansion and provided the means of improving the town's trade after a period of depression.

Although there are references to the town's earlier history, necessary in order to set the scene for the story of the port in the 19th and 20th centuries, this book concentrates on the story of the dock itself, from its construction in the 1830s to the present. The dock was important not only to Ipswich but to a much wider area of Suffolk; it is for this reason that mention is made of the Stowmarket Navigation, whose own history is inexorably linked with that of the dock.

The names of many of today's quays evoke memories of merchants long dead, and of the cargoes that were once landed from the ships that berthed at those quays. Those names, and the names of the many dockside taverns that existed a hundred years ago, form a rich pattern which the writers, former Ipswich Port Authority personnel manager Bob Jones and local historian Bob Malster, have attempted to weave into this book. A feature of that tapestry is the old Promenade—the tree-lined thoroughfare that before the First World War provided the people of Ipswich with a pleasant Sunday afternoon stroll and a vantage point from which to view the shipping in the dock.

Today the dock is a very different place. Sailing ships and barges have given way very largely to powered craft, and the only sailing vessels to be seen are pleasure craft.

With the development of the port's operations downstream of the dock, able to accept vessels of up to 18,000 GRT at nearly all states of the tide, the Wet Dock now handles only a small proportion of the five million tonnes of cargo to pass annually through the port.

Whatever the future of the dock and the surrounding estate, the 150-year story of the Wet Dock, short though it is when compared with the 2,000-year history of the port, is one that deserves to be told.

Early Days 1

As THE 18th century drew to a close, Ipswich was beginning to awake from a period of depression that had seen grass growing in the streets and mud banks accumulating in the river. During the 850 years that had passed since the Danes had invaded East Anglia and occupied the Anglo-Saxon settlement of Gippeswyk, the town had experienced alternate periods of prosperity and stagnation; and there has always been a close link between the fluctuations in the town's fortunes and the state of the river.

Situated eight miles from the sea at the head of a shallow estuary, Ipswich had never been an easy port to approach. The Orwell, whilst flooding over the mudflats at high water to present an appearance of a wide and safe river, shrank at low tide into a narrow, meandering channel that was difficult to navigate, even for small sailing vessels. Acute bends hindered ships seeking to follow the channel, and at times shoals prevented larger vessels from reaching the town quays.

An inbound vessel, having negotiated the tortuous passage past such quaintly-named banks as Mulberry Middle and through aptly-named channels like Back-Again Reach, would pass Hog Highland (where Cliff Quay now stands) before winding into the commercial heart of the town. On the south side of the river was a wide tract of marshy land bordered by mudflats that became visible at low tide, but the north bank was very different. Here merchants' houses and their gardens came down to the water's edge, and there were warehouses and quays at which vessels could load and unload their cargoes.

Some of these were private quays built by merchants for their own use, but there was also a Common Quay that was provided for the common use of the town's merchants. Such a quay was probably in existence in the 13th century, when the foundations of the town's local government were being laid.

The impecunious King John, not very long after his accession in 1199, granted Ipswich its first charter, allowing it the privilege of ordering its own affairs. Appropriately enough, a sailing ship featured on the Common Seal made in 1200. It is the earliest known representation of a ship with a rudder hung on the sternpost, instead of the steering oar over the quarter that had been used previously.

Records show that in 1280 the port's trade warranted the appointment of a Collector of Customs, and in 1404 and again in 1464 Ipswich was appointed a staple port, one of very few ports through which wool could be exported to the Continent.

The end of the 16th century was a time of great prosperity for Ipswich and its inhabitants. At that time, according to an 18th

Opposite page:
Ships lying at quays in the Orwell below Stoke Bridge, as portrayed in an engraving by C. Pyke dated 1801.
Ipswich Borough Museums and Galleries

century author, Ipswich had "a great trade, and was very populous, having an abundance of rich merchants in it." In the century that followed, however, the trade in cloth declined dramatically and, despite a minor resurgence in the grain trade, the port and town suffered a lengthy depression. The same author continued that the town "at present is much decayed in its trade and people."

As might be expected at such a period of depression, the maintenance of the quays and of the river itself was neglected during the 18th century, and at low water the river almost dried out by the Common Quay. Vessels drawing more than eight feet were unable to reach the town's quays and had to unload their cargoes into lighters at Downham Reach, three miles downstream of Ipswich. The mud that was choking the Orwell threatened to throttle the town's trade.

By the end of the 18th century, however, there were signs of an upturn in the town's fortunes, and the state of the river was not to be allowed to hold back that improvement in trade being brought about by new industries and a new spirit of enterprise. Thus it was that disgruntled merchants and businessmen sought the advice of William Chapman, an eminent engineer from Newcastle-upon-Tyne, who had been much concerned with inland waterways and other projects.

Chapman made his first report in 1797, and over the next eight years he provided a variety of proposals which proved too ambitious for the Ipswich men. His first reports proposed a ship canal from Downham Reach up to a "proposed bason or floating dock" eleven acres in extent or, alternatively, a new channel that would by-pass such meanderings as Back-Again Reach.

Chapman's ideas were likely to prove costly in implementation, and he was sent away to think again. Nevertheless, those ideas were to be catalysts for subsequent developments.

Activity on the Common Quay in the early 1800s. This was the only quay in the town at which any ship might load or unload, the others being privately owned by merchants.
Robert Malster

The River Commissioners 2

BY THE beginning of the 19th century the "Committee of Subscribers for the Improvement of the Port of Ipswich", to which William Chapman submitted his several reports, could put off no longer the action required to halt the town's decline. Whereas in 1755 there were 12,124 inhabitants, in 1801 the population had fallen to 11,061, and the town's new industries were held back by the inadequate state of the Orwell.

On 15 February 1803, a general meeting of the inhabitants of Ipswich was held at the Moot Hall on Cornhill to consider what ought to be done to improve the situation, and it was decided to seek Chapman's advice once more. Making his third report in 1804, Chapman submitted two plans for deepening the river and straightening it to enable vessels of 250 tons to reach the quay at spring tides.

He felt it necessary to point out, however, that although this plan would cost a mere £15,000, there would be a need for continuing expenditure on dredging, "of which the advantage of ballasting the ships will form but a small part". Here he was referring to the monopoly held by the Corporation of supplying ballast to vessels leaving without cargo.

His frustration was evident. It really burst forth when he penned a postscript to his report after it had been read to a meeting of the committee on the eve of another general meeting of the inhabitants on 13 November 1803.

> Gents,
> From the bent of the Public opinion appearing to me to be in favor of the expedient of simply deepening the River and Cutting off some of the bends because of the idea that it will be adequate to all the wants of the town and that the superior advantages of the plan recommended would in the general opinion not compensate for the greater estimated outlay . . . it appears to me clearly advisable not to bring forward my report to the General Meeting tomorrow, because by creating a secession from those who are in favor of the only plan likely to be carried into effect, it may prevent anything whatever being done.

The inhabitants and the Corporation both set up committees to consider a Bill that should be presented to Parliament. In an attempt to gain control of the income from shipping dues the Corporation proposed that all 41 of their members should be River Commissioners, leaving only 25 commissioners to be appointed on behalf of the inhabitants. Such a proposal did not meet with the approval of men who were convinced that the Corporation could not be relied on to carry the proposed Act into effect if the powers were vested in them. So the Committee of Subscribers for the Improvement of the Port of Ipswich went ahead on its own and drew up a Bill that was laid before Parliament.

The comments of this committee on the Corporation and its behaviour left no doubt in anyone's mind of their opinion of that organisation's past record. "It is true that they are the supposed conservators of the soil of the river, and it is admitted that they have faithfully preserved it from injury by not suffering it to be removed for the accommodation of the trade and shipping, for want of which the evils now complained of have principally arisen," the committee said in a memorandum.

Not surprisingly, several of the clauses in the Bill drawn up by the Committee of Subscribers were designed to affect the rights and interests of the Corporation. Equally unsurprisingly, the Corporation saw these clauses as a threat to their control of the port and river, and to the income received from the town's trade. When the Corporation petitioned the House of Commons against the Bill "in order to get those objectionable clauses expunged or amended", however, their own record of inactivity told against them.

Simon Jackaman, clerk to the Ipswich River Commissioners, 1805-1837.
Ipswich Port Authority

The Act for "improving and rendering more commodious the Port of Ipswich in the County of Suffolk" went through all its stages in Parliament and received Royal Assent from King George III in 1805. The "interested gentlemen" who had promoted the new legislation were then enabled to come together formally as River Commissioners who would act effectively as guardians of the port.

The first meeting of the River Commissioners was held at Gooding's Coffee House on the corner of Tavern Street and Tower Street (on the site where British Home Stores now stands) on 23 July 1805. At that meeting 49 of the 72 commissioners were sworn in, and they appointed Simon Jackaman their first clerk and solicitor at an annual salary of £20, plus fees.

The first Harbourmaster, Robert Cole, was appointed in August and an assistant in the following month. It was also agreed that 12 pilots would be needed to guide ships through the tortuous upper reaches of the river.

There was an enormous amount of work to be done to remedy long-term neglect of the Orwell and of the town's waterfront. The commissioners met weekly, and quickly began to get through a great deal of vital business.

By far the most important task that confronted the commissioners was the clearing of the accumulations of mud that were blocking the channel. In September they agreed to purchase a "Steam Dredging Engine", one of the first steam dredgers to be put to work anywhere in Britain, a fact of which the commissioners were justly proud. It was "to be ready by Christmas"; in fact the dredger seems not to have begun work until the following July, and even then it suffered from some initial teething problems.

Gooding's Coffee House in Tavern Street, where the River Commissioners held their first meeting in 1805.
Robert Malster

Before the steam dredger could be set to work, some manual removal of the mud was needed. The Harbourmaster was ordered to "engage any number of men he can to excavate the mud in the intended new cut between the Carpenters Quays and the Cliff to such a depth as may be necessary to enable the Steam Engine Dredge Barge to work at half flood tide." Anyone who has attempted to walk out on the mudflats of the Orwell will know what a perilous and filthy task the "Mud Men" had to perform as they cut their way through the bank to prepare the way for the dredger. At the end of the day's work they must have fully lived up to the name given them.

The River Commissioners' first full year of operations resulted in a small loss, the dues on goods and tonnage on vessels amounting to a little over £3,000. However, more than half the outgoings of over

£3,090 were accounted for by costs of obtaining the Act of Parliament. Money to invest in new equipment, such as the steam dredger, had to be obtained by subscription, many of the commissioners putting up their own cash for this purpose. One of the most munificent was Samuel Alexander, the banker, who subscribed no less than £3,150 during the first year, more than enough to purchase the dredger.

Whilst some of the commissioners were very ready to provide the cash necessary for the work of improving the river, some strange financial practices still appear to have existed. To give just one

A plan showing the improvements proposed in 1804 and eventually put into effect by the River Commissioners.
Ipswich Port Authority

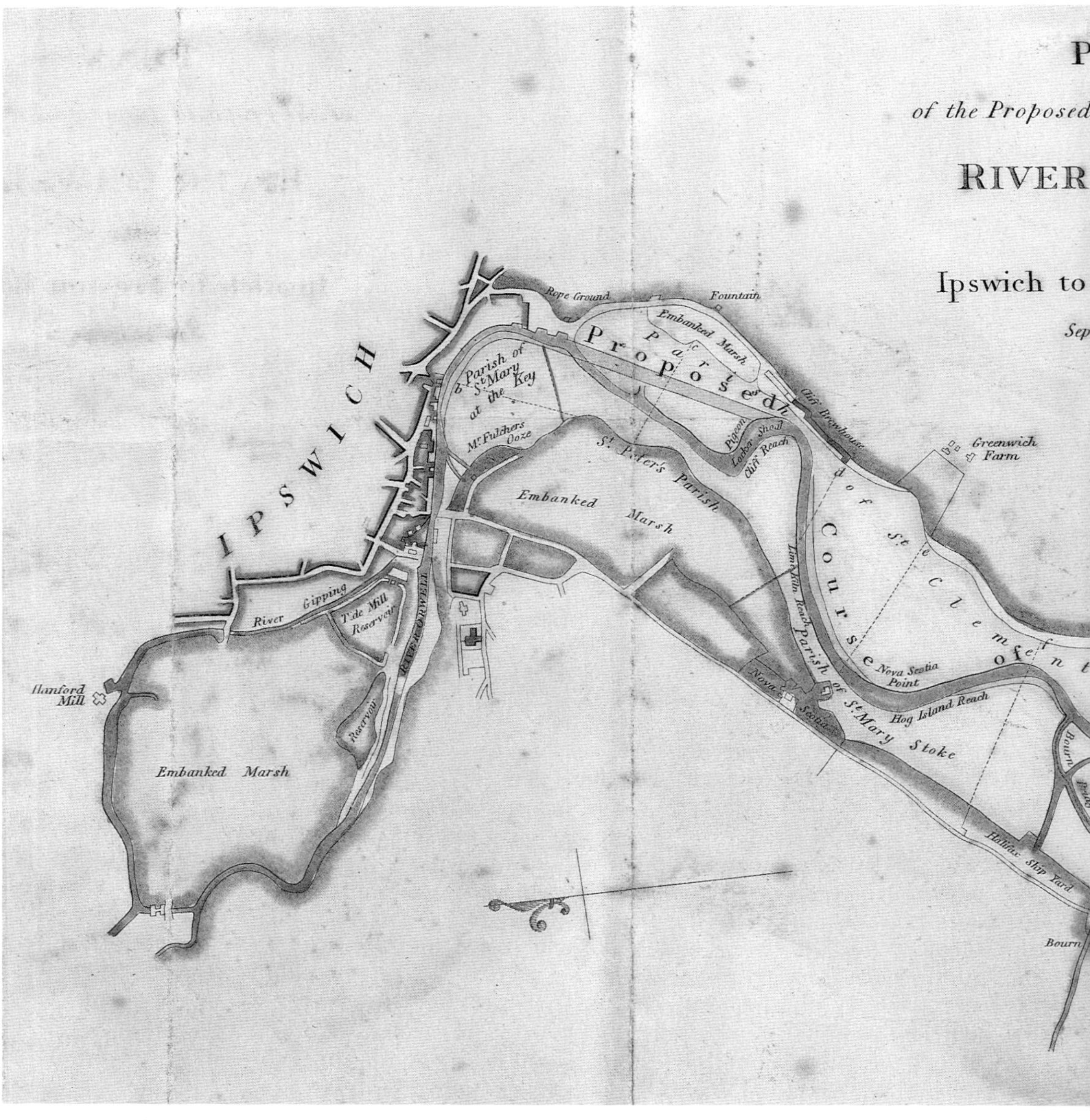

example, Robert Cole resigned as Harbourmaster in February 1807, just a year and a half after his appointment, because he was fed up with having to pay the labourers' wages out of his own pocket, and then having to wait for the commissioners to repay him at the end of each month. Within weeks he was replaced by Miles Davis.

The loss of the first year was turned into profit in the second and third years of trading, £473 and £740 respectively. The steam dredger appears to have been worth its expense, for in 1808 it was calculated that 26,166 tons of soil had been raised during the past year at "less than half what it could cost by manual labour." A report

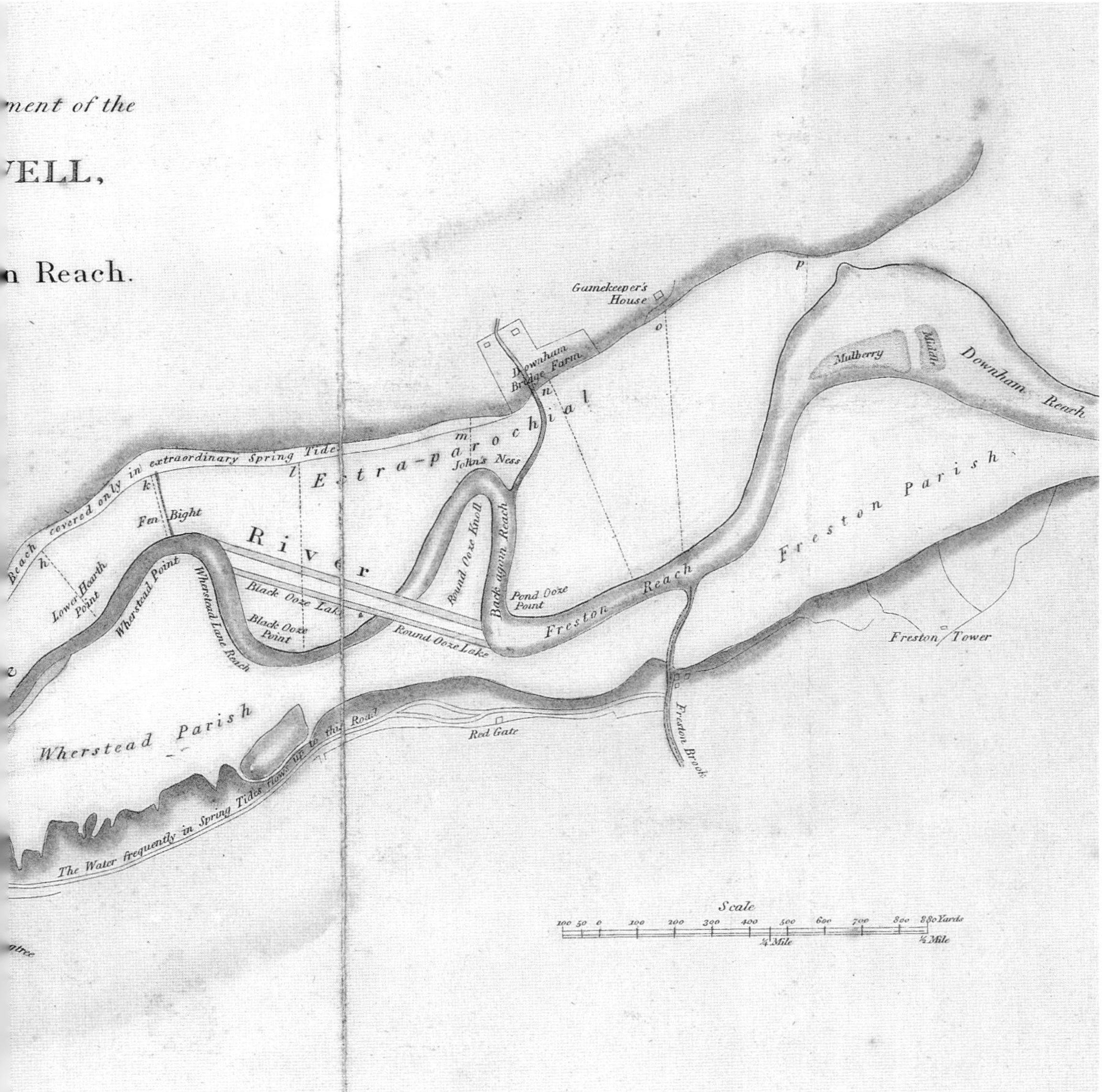

to the commissioners by their engineer, William Shaw, estimated that the Dredging Engine would defray its costs and charges "in a few years."

Sadly, this was not to be the case. In the following years the dredger had to be laid up for lengthy periods each winter for maintenance, and the necessary repairs and replacement parts proved costly. More reliance had to be placed on the relatively ineffective method of clearing the mud by hand. In 1812, indeed, the dredger was used for only two months of the year, and no less than £583 had to be spent in wages for excavating the soil. "Only so much labour is expended as is unavoidable to maintain a free navigation of the River from growing shoals", said Shaw in his report for that year.

The Commissioners worked valiantly, raising money where it was needed and spending it as efficiently as possible to keep the port in operation, but their problems were not confined to the river itself. In 1814 a survey of the Common Quay revealed that it was in a very bad state of repair, and the commissioners were obliged to order that the steam dredger should not work near it "to prevent its falling in." Eleven years later, it was reported that vessels were unable to

Ipswich from the Cliff, a lithograph by L. Hague after a painting by the Rev. Richard Cobbold. The River Commissioners' steam dredger can be seen at work. Ipswich Borough Museums and Galleries

berth alongside the Common Quay because of the accumulation of soil on the river bed.

The repaired dredger performed miracles in the river, excavating more than 35,000 tons in 1817 alone. In addition to this several thousand tons were cleared by manual labour, "where the Engine could not be advantageously employed." Such spoil as was suitable for use as ballast—and to judge from the complaints, much that was not— was deposited at the Ballast Wharf for sale to colliers and other vessels sailing in ballast. What was clearly unsuitable for this purpose was dumped either at Redgate or in John's Ness Hole, a mile or so downriver from the town's quays.

Old-time sailing ships needed to be loaded in order to sail from port to port, and if they did not have a ready cargo it was necessary for them to load ballast to bring them down to sailing trim. Some were so unstable when unloaded that they could not even move from one berth to another without some form of ballast to keep them upright.

The necessary ballast was usually supplied in the form of sand or gravel; the term "ballast" for material used in the building and construction industry is derived from its use for ballasting ships. At Ipswich, the gravel supplied at the ballast wharf, situated on the east side of the river where Ransomes' Orwell Works later stood, was mainly dredged from the river. The old Corporation, which owned the bed of the River Orwell, held a monopoly in the supply of ballast and carried on a most lucrative trade, the income from which disappeared into the Corporation's coffers and was not employed to improve the river. It was because of the Corporation's long-term record of neglect that, in 1805, the men behind the promotion of the Act of Parliament setting up the River Commissioners insisted that the new body should take over the responsibility for supplying ballast to shipping, applying the proceeds to improving the navigation of the river.

Much of the ballast supplied at Ipswich went into the holds of the sailing colliers which carried on such an extensive trade with the North-east coast ports. The ballast heaps on the banks of the Tyne were once a familiar sight to those sailing into that river of coal.

With the construction of the dock, the ballast wharf was moved to a new site on the river side of the lower embankment. Nineteenth-century maps and plans show the wharf projecting from the embankment to provide a quay at which ships could lie to load ballast, landed on the wharf from the lighters built for the Commissioners by one of the Ipswich shipbuilders to carry the material from the steam dredger. It was fortunate for the Commissioners that so much of the material taken from the river when cutting new channels and deepening the old ones was so suitable for use as ballast. There were, however, occasional complaints that the material landed contained too much mud and was liable to shift during the voyage—shifting cargo or ballast could easily result in the loss of a vessel.

Disaster struck in 1818 when a severe April flood destroyed Stoke Bridge, a neat structure of stone connecting Ipswich with the parish of Stoke and the only bridge over the tidal Orwell within the borough; Princes Street bridge was not built until 1849. The commissioners quickly arranged for six of their punts to be lashed together with a decking across to form a temporary floating bridge, enabling people and carriages to cross the river. Amazingly, this contraption held together until September 1819, when a replacement bridge built by Ransome and Sons under the superintendence of William Cubitt was opened to the public.

By 1821 the town's population had risen dramatically to 17,156, an increase of 70% in 20 years, largely as a result of an influx of people from the surrounding countryside coming to work in Ransomes' factory and in the other industrial concerns that were growing up in the town. Ipswich was clearly an expanding town, and that expansion was to continue throughout the 19th and 20th centuries.

The cast-iron bridge that replaced the earlier Stoke Bridge in 1819, seen in a photograph taken in the 1850s by Robert Burrows.
Suffolk Record Office

Right: *The Wet Dock seen from Stoke Bridge in a painting by Claude Lorraine Nursey. At extreme left is the building of the Ipswich Steam Navigation Company, which operated steamers to London.*
Ipswich Borough Museums and Galleries

Below: *The Ipswich quays in 1753, seen in a detail from John Cleveley's panoramic painting of Ipswich.*
Ipswich Borough Museums and Galleries

Whilst the port's trade was basically coastal, links were established with Holland, Germany and the Baltic - links that remain unbroken today. The coastal trade was principally with London, which was supplied with both grain and flour, as well as with other products of the Suffolk farms. Corn and flour were also sent to Liverpool, from where salt was brought back. There were then no railways to carry such goods across England, and coastal shipping was the only satisfactory way of distributing goods. In the mid-1820s, Ipswich was importing between 12,000 and 15,000 tons of coal per annum, while sending between 17,000 and 23,000 tons of wheat to London each year.

The vessels in which these cargoes were carried were mainly small, but they still required pilots to bring them up the river. Several of the pilots found themselves in trouble with the commissioners, mainly through drinking rather too much. In 1824 Thomas Downs, having been suspended for "misbehaviour", was so unwise as to appear before the commissioners in a state of intoxication, and was suspended for a further week. He was eventually reinstated following recommendations from several ships' masters. Three years later the rules were clearly laid down: "If any Pilot . . . appointed to take charge of any vessel within this port shall go on board such vessel in a state of intoxication . . . the Harbourmaster is required to suspend such Pilot . . . until the next General Meeting of the Commissioners."

In 1835, the pilots were ordered to navigate all vessels out of the port as far as Pin Mill, for the same rates that they had been charging for taking ships to Downham Reach. The pilots protested at this ruling, and they and the commissioners eventually came to a compromise. The pilots provided, at their own expense, another pilot for the additional part of the journey, and agreed that if this person were not available they would do the extra voyage themselves at no extra charge.

By 1835, the River Commissioners had achieved a great deal, against all odds. They had generally managed to keep the river clear of obstructions, trade had continued at a reasonable pace, and they had managed to raise sufficient funds to pay their employees. They had in 1820 made a cut through Round Ooze, making it unnecessary for vessels to use the circuitous and difficult channels by John's Ness, and the following year they had begun a further cut to take ships straight across the bank between Hog Island Reach and Limekiln Reach.

While achieving all this, they had by 1830 paid off the original loan of £8,000, borrowed on security of the rates to be levied on goods entering the port. They also managed to accumulate a surplus of £25,000 that would eventually be passed on to their successors to help finance the construction of the Wet Dock.

There were among the River Commissioners men who believed that all this was not enough, and that more should be done to improve the port, to make it suitable not just for the coasting trade but for foreign trade as well. They doubtless remembered William

Opposite page: *The old Custom House and the crane on the Common Quay before the building of the dock, from a painting which now hangs in the offices of the Ipswich Port Authority in the Old Custom House.*
Ipswich Port Authority

Opposite page:
Frederick Brett Russel's watercolour of the Custom House designed by John Medland Clark and opened in 1845. It housed the office of the Ipswich Dock Commission as well as the Customs and Excise offices.
Ipswich Borough Museums and Galleries

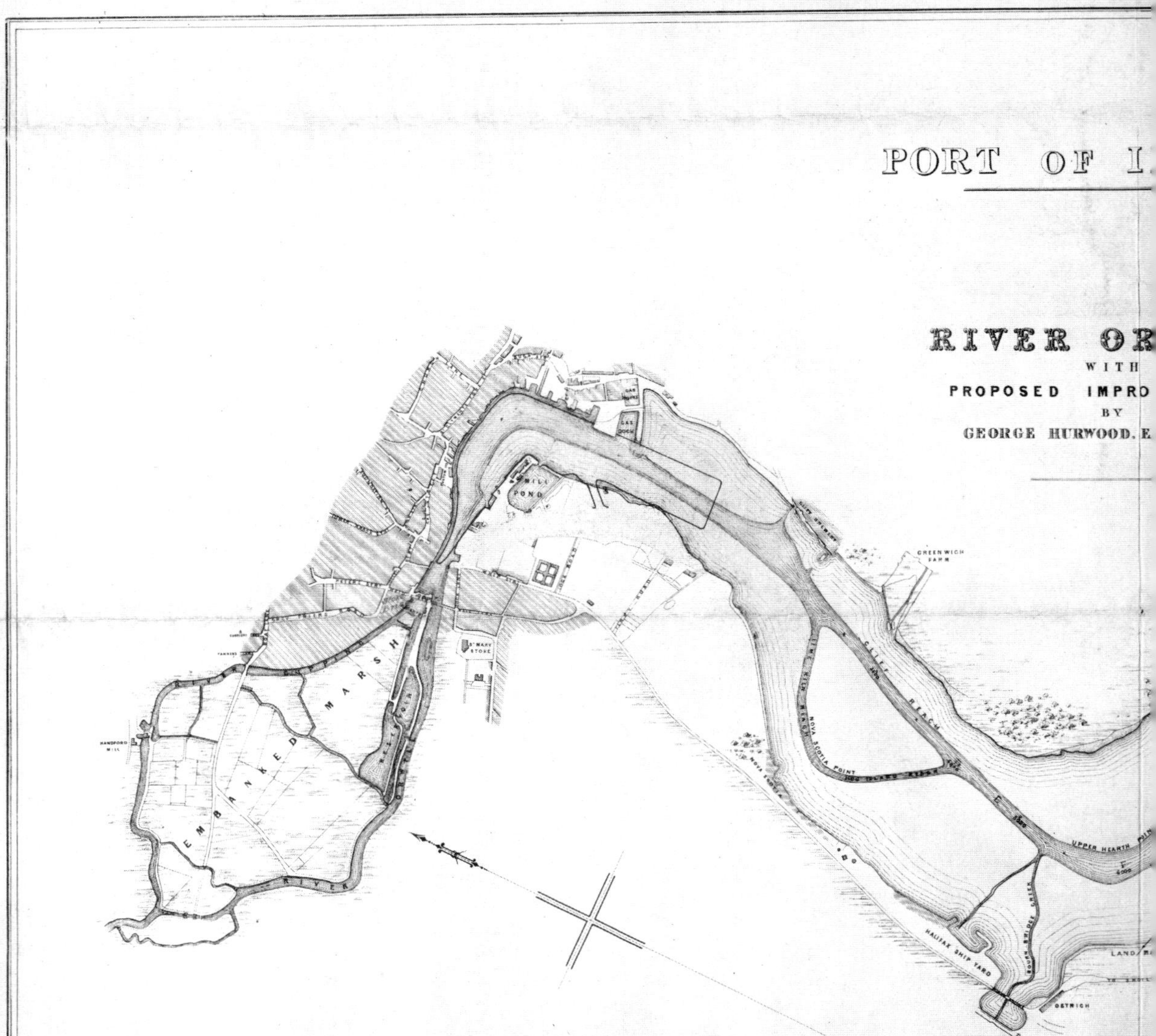

N.B. The Red line denotes the site of the Wet Dock

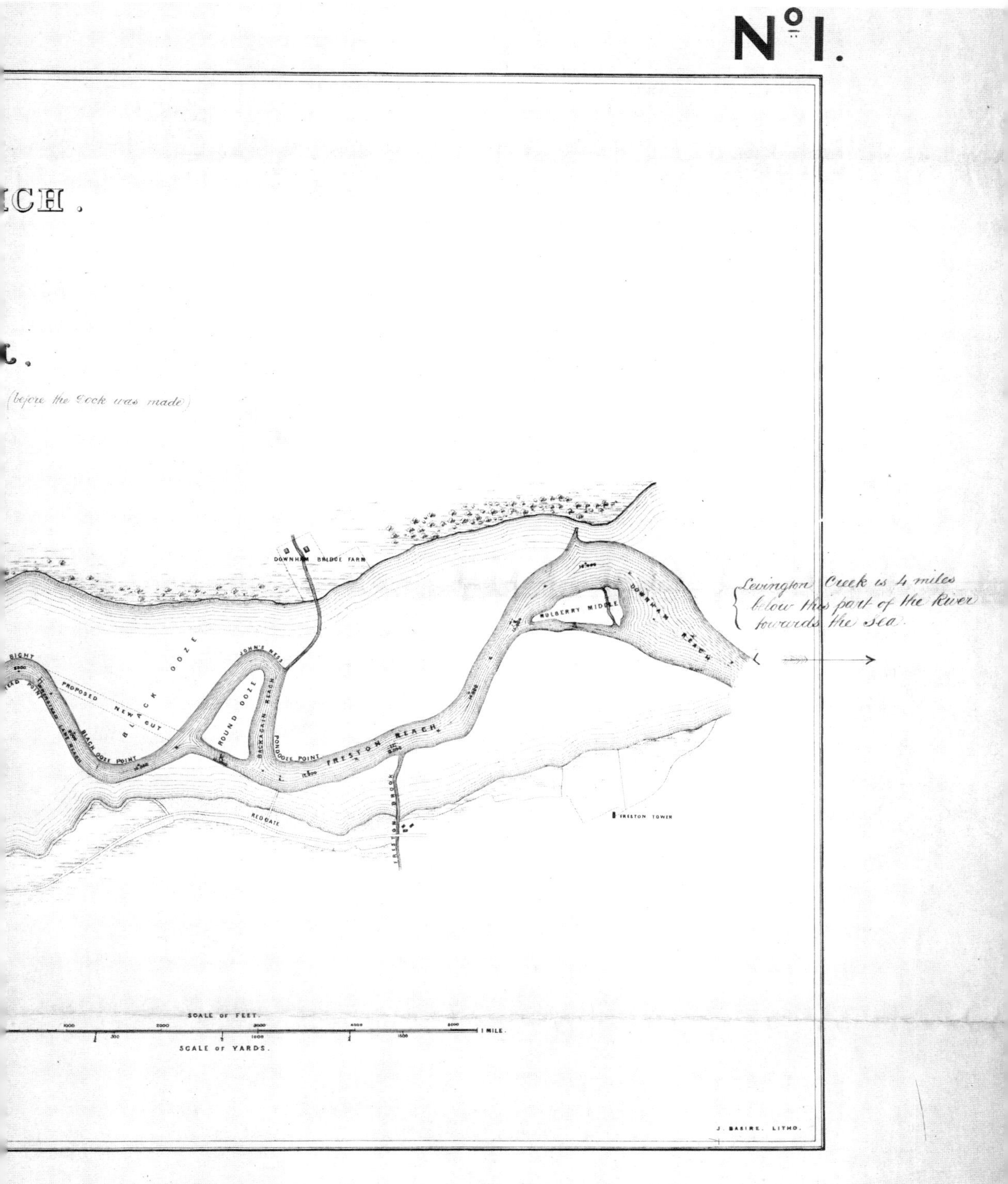
Nº I.
ICH.
(before the Dock was made)
DOWNHAM BRIDGE FARM
MULBERRY MIDDLE
DOWNHAM REACH
Levington Creek is 4 miles
below this part of the River
towards the Sea
BLACK OOZE
JOHN'S NESS
PROPOSED NEW CUT
ROUND OOZE
BLACK OOZE POINT
POND OOZE POINT
FRESTON REACH
REDGATE
FRESTON BROOK
FRESTON TOWER
SCALE OF FEET
1 MILE
SCALE OF YARDS
J. BASIRE. LITHO.

Chapman's exhortations that the town would eventually regret not having carried out his original plan for a ship canal and dock.

If Chapman had not died in 1832, it is very likely that he would have been at the meeting held in the Moot Hall on 19 January 1836, under the chairmanship of the banker Dykes Alexander, "to take into consideration a proposal for forming a Wet Dock at this Port, and improving the River and Harbour".

William Lane, the Collector of Customs for the port, and in earlier years a Bailiff of the town, spoke at the meeting of a plan he had prepared for forming the existing channel of the river from St Peter's Dock to the Ballast Wharf into a wet dock, and bypassing it by a cut to carry the tidal water. Lane estimated that the work would cost between £25,000 and £30,000, and he suggested that an Act of Parliament should be sought, allowing the £25,000 accumulated by the River Commissioners to be used for this purpose.

William Lane is thus seen as the man who conceived the idea of a wet dock formed out of a bend in the river, though in suggesting this he was clearly drawing on the earlier plans of William Chapman. When, three years later, Mayor George Green Sampson laid the foundation stone of the lock, he called for "three times three" cheers for one "with whom the very germ of the work originated, who was influenced by no private motives, but strictly for the public good — William Lane, Esq."

A committee was appointed at the public meeting to collect information on the proposal, and this "Dock Committee", which included among its members a number of River Commissioners, sought the help of a prominent London engineer, Henry Palmer, a vice-president of the Institution of Civil Engineers. He was asked to report on the viability of a wet dock in which ships could lie afloat at all stages of the tide.

Palmer's plans came up for discussion at a public meeting in the Town Hall on 4 November 1836. Despite his absence from the meeting—he was indisposed at the Great White Horse Hotel, having travelled all the way from Wales to attend the meeting—the assembly expressed itself in favour of his "voluminous and excellent report". His involuntary absence was unfortunate, and might well have set some of those concerned with the proposals against him; in subsequent days there were complaints that he had treated the Dock Commissioners with contempt by being absent from their monthly meetings.

The scene was at last set for progress to be made. William Chapman would certainly have applauded those who fought for new improvements, for in a report of November 1804 he had stressed the need for boldness. Pointing out the great expense of using the West India Docks and other facilities in London, he had urged the Ipswich merchants to provide the kind of facilities that would attract trade away from the Thames to the Orwell, and had commented that "the time may shortly come when the subscribers may find the disadvantage of their works not having been executed on a large scale".

Previous pages: *A plan by George Hurwood showing the Orwell as it was before the construction of the dock. This plan illustrated Hurwood's lecture to the Institution of Civil Engineers on the construction of the dock, seen marked in outline.* Ipswich Port Authority

The Building of the Dock 3

PALMER'S proposals to the Dock Committee were certainly ambitious, with a dock of 33 acres water area that would be the biggest in Britain; none of the docks on the Thames at that time approached the size of that proposed for Ipswich, and those at Hull and Liverpool were dwarfed by it.

The Dock Committee went ahead with the preparation of a Parliamentary Bill that would enable a new body of commissioners to raise money for the building of a dock, and it really seemed that Ipswich was to be given the facilities it so much needed. At the eleventh hour a diehard group of River Commissioners, headed by Welham Clarke, proposed at a meeting of the Commission that they should pay out of their funds "all expenses the gentlemen of the Dock Committee had incurred up to the present time, provided they throw the Dock Bill overboard into Deep Water".

Put to a meeting of the River Commissioners on 19 June 1837, this proposal was convincingly defeated. That proved to be the River Commissioners' final meeting. On 30 June that same year, just ten days after Queen Victoria's accession to the throne, the Ipswich Dock Act received Royal Assent; the future of the port lay with a new body of Dock Commissioners.

In his first report, Henry Palmer had referred to three plans which, he said, required special notice. One was Chapman's for a ship canal and dock; the second was a suggestion that had been made by a certain J. Hare, and put at the initial public meeting by Dykes Alexander, for the building of an embankment near Downham Reach which would have made the whole river above that point into an immense non-tidal dock; and the third was William Lane's. The scheme that he submitted owed much to both Chapman and Lane; he dismissed the idea of a dam across the river as being impractical.

Palmer worked hard on his drawings, and by September 1837 was able to report to the Commissioners that he hoped soon to complete all plans and specifications. A month later, on 23 October, the plans were ready for inspection. Tenders were invited for the construction of the dock, and the lowest tender of £65,178 from David Thornbory, of King's Lynn, was accepted in March 1838.

Already opposition to the scheme was beginning to grow in Ipswich. That opposition was led by one of the local newspapers, the *Ipswich Journal*, which printed readers' letters agreeing with its editorial worries about frost damage to quay walls and the effect of high winds on vessels entering or leaving the lock. Another newspaper, the *Ipswich Express*, with other political affiliations, expressed itself very much in favour of the dock scheme and reported progress enthusiastically.

The Commissioners themselves were very much concerned about the costs involved. In April 1838, they indicated that the amount of the proposed contract with Mr Thornbory "approaches very nearly to the means at the disposal of the Commissioners", and wondered if it might be "expedient to modify the specifications so as to reduce the amount of the outlay without impairing the general effect of the works."

Following a conference between the Commissioners, Palmer and Thornbory, the Management Committee recommended that the Commissioners "limit the extent of the Dock in the first instance to the Ballast Quay". The Commissioners accepted this recommendation in May 1838, and a month later, on 12 June, the contract with David Thornbory was signed.

Digging had begun three days earlier, those employed on the task being "fifty labourers residing in the neighbourhood of the town hitherto in search of employment". The Mayor, attorney Peter Bartholomew Long, who was also Clerk to the Dock Commissioners, laid the first stone of the new quays on 13 August.

Peter Bartholomew Long, clerk to Ipswich Dock Commissioners and three times Mayor of Ipswich.
Ipswich Port Authority

The general specifications of the works to be executed were quite detailed. The main part of the work was the construction of a "Wet Dock or Basin, in which Vessels can at all times lie afloat, the said Dock or Basin being made to take the place of a circuity of the River Orwell, the Water being constantly penned up, or sustained to a sufficient depth by means of Dams or Embankments constructed across the course of the River, at the upper and lower ends of the space which is to become the dock".

Then there were specifications for the cutting of a new channel to take the water of the River Gipping on its way to the sea; for the building of the lock through which ships could gain access to the dock; and for the erection of a quay along the north side of the dock so that people and goods could for the first time pass along the waterside. There were also specifications for the roadways alongside the dock and the new river channel. There was to be a sewer discharging into the Orwell to carry away the town's drainage, which had formerly discharged into the section of river that was to constitute the dock.

The requirements for the two dams at the upper and lower ends of the dock were particularly detailed, as indeed they needed to be. "The Embankment at the lower end of the Dock is to be composed of Gravel and Clay uniformly mixed. Within the Bank . . . a Wall of Puddle, composed of one part Chalk or Clay to 5 parts Gravel and Sand or other materials equally suitable. It is to be formed 10 feet in thickness; it is to be sunk two feet below the natural bed of the River and to extend one foot from the surface of the Bank, the whole of the other part of the Bank is to be well founded or otherwise worked so as to form one compact solid mass."

The conditions for the construction of the quays were equally precise. "The Wall is to be built in parallel courses, the bricks to be used in it are to be the best hard Grey Stocks well burnt so that the sand contained in them shall become vitrified. The Mortar to be used is to be composed of Blue Lias Lime or of other stone Lime of good quality which will set well under water and sharp clean Sand in the proportion of one part Lime to three parts Sand. The Lime is to be well burnt and slacked with water and to be well worked with the sand in a rolling mill until an equal and uniform mixture is formed."

Once begun, the work continued apace. A year later, on 26 June 1839, the foundation stone of the lock was laid by the new Mayor, George Green Sampson. In those days such civic entertainment attracted an enormous following; and on this very special day some 15,000 people turned out to watch the Mayor process with a bodyguard of police, and accompanied by aldermen, councillors and harbour officials from the Town Hall to the riverside excavations. It was a hot summer's day; and the band of the 9th Lancers, their tunics faced with scarlet and blue, played at the dockside to entertain those who awaited the Mayor's arrival through

Henry Davy's print of the laying of the first stone of the lock by George Green Sampson in 1839.
Ipswich Port Authority

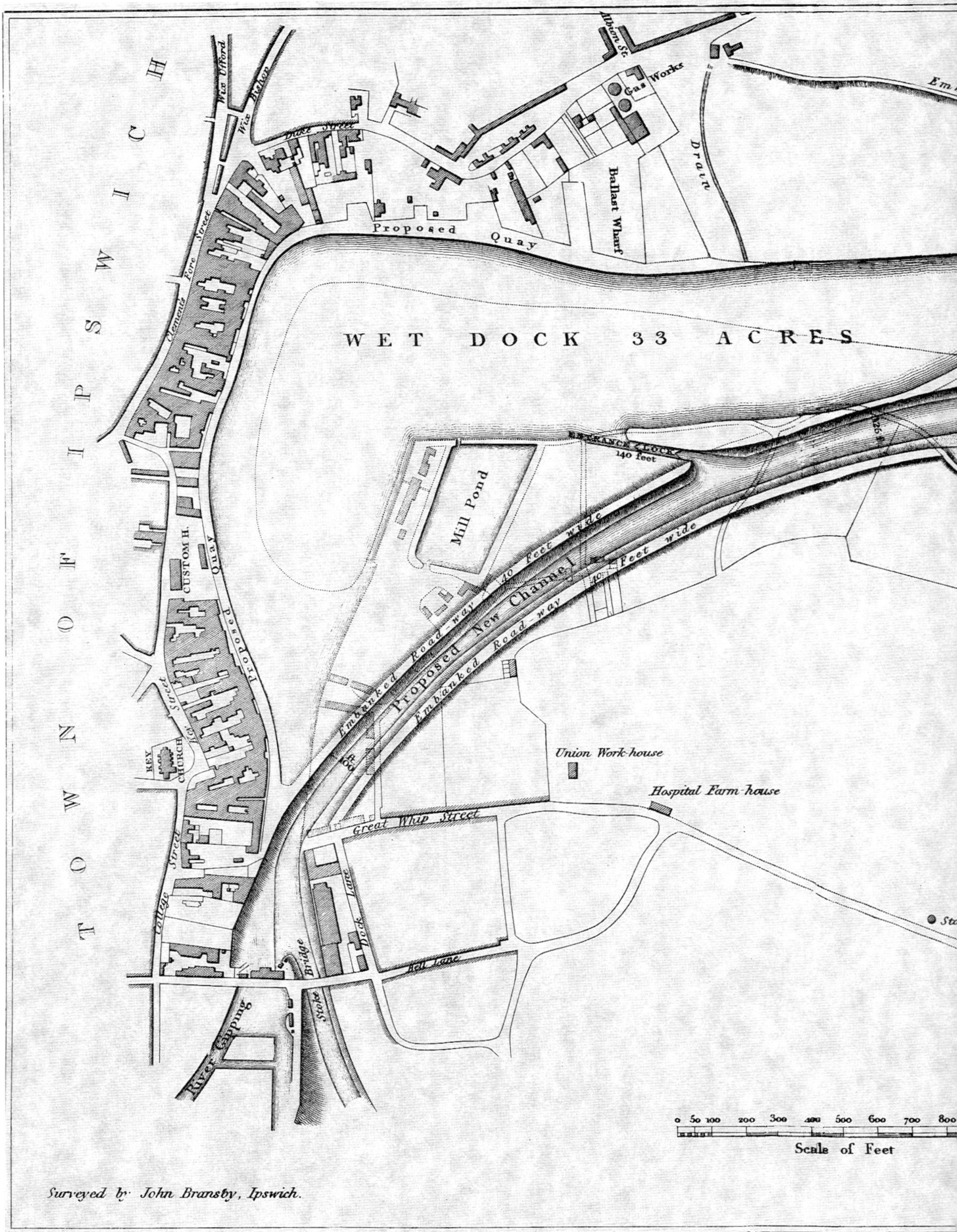
TOWN OF IPSWICH
WET DOCK 33 ACRES
Proposed Quay
Proposed New Channel
Embanked Road-way
40 Feet wide
Mill Pond
ENTRANCE LOCK
140 feet
Gas Works
Ballast Wharf
Duke Street
Albion St.
CUSTOM H.
KEY CHURCH
Key Street
College Street
Clements Fore Street
Union Work-house
Hospital Farm-house
Great Whip Street
Bell Lane
Dock Lane
Stoke Bridge
River Gipping
Scale of Feet
0 50 100 200 300 400 500 600 700 800
Surveyed by: John Bransby, Ipswich.

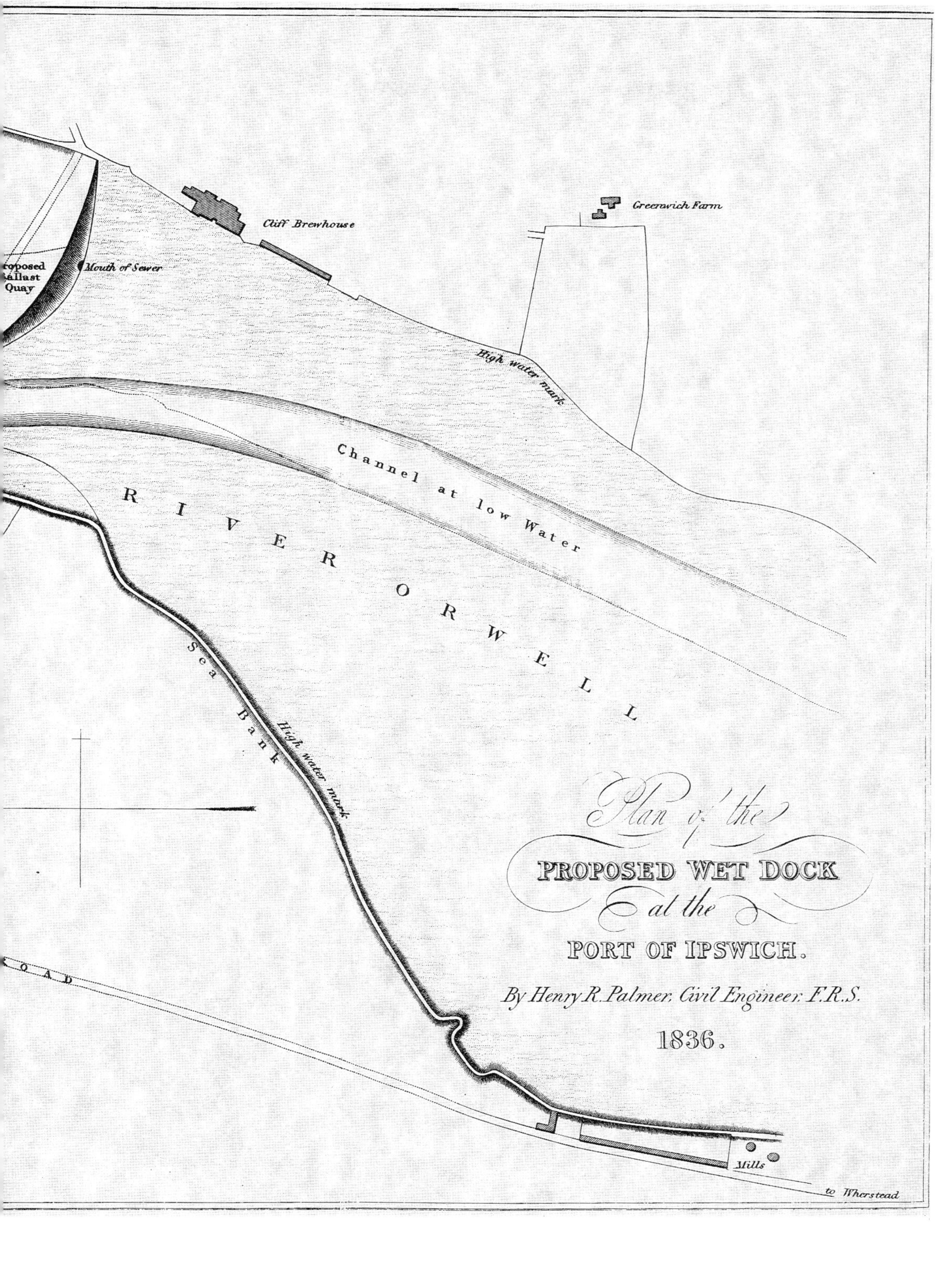
Cliff Brewhouse
Greenwich Farm
roposed
Ballast
Quay
Mouth of Sewer
High water mark
Channel at low Water
RIVER ORWELL
Sea Bank
High water mark
Plan of the
PROPOSED WET DOCK
at the
PORT OF IPSWICH.
By Henry R. Palmer, Civil Engineer, F.R.S.
1836.
Mills
to Wherstead

a triumphal arch crowned with a model schooner and the motto, "May Ipswich Flourish". Cannons boomed out, and there was loud cheering from the masses as the civic procession appeared. Mr Thornbory was presented with a purse of money, presumably a token gesture, and the townspeople were treated to a grand firework display.

No doubt the Dock Commissioners felt that they were justified in spending a relatively small amount on such celebrations. Still to be seen in the Port Authority archives are "Bills re Laying First Stone of Lock". Prominent among the items are "Gunpowder £2 1s. 6d" and "To Carting 2 Cannons to and from Ballast Wharf—3s—".

With the foundation stone in place, the building of the lock went ahead without delay; and on 1 October 1839 David Thornbory provided a dinner for his workmen "at the bottom of the newly-erected Dock entrance" to celebrate the completion of this part of the works.

"The men were seated at a table extending the whole length of the entrance, having only room for a cross table at the top, at which Mr Thornbory presided, surrounded by several ladies and gentlemen, who were politely invited to partake of his hospitality . . . ", while "several hundreds of spectators looked down upon the joyous scene from an elevation of many feet . . . "

The protection of the works from thieves and vandals cannot have been an easy task, for there was no way of preventing access to the scene of operations. In January 1840, Mr Thornbory "presented the police with £10 as an acknowledgement of the protection exercised over the property constantly exposed at the works". Later the Dock Commissioners resolved that "inasmuch as the duties of the Police will be materially increased by the additional watching necessary at the quay, the Commissioners suggest to the Watch Committee the propriety of adding one man to the established force . . . the expense of which addition the Commissioners will undertake to defray".

By the early months of 1840 the New Cut was almost complete, although it was still separated from the river by embankments at each end; and the lock needed only its gates, while some 2,000 feet of the new quay on the north side of what was to be the dock had been completed. "Not the smallest settlement has taken place in the whole length," stated the *Ipswich Journal*.

This was, however, the only good thing the newspaper could find to say about the project. It continued its tirade against the dock, hinting in May 1840 at "a vast increase of expenditure". Indeed there was; the following year the Commissioners had to obtain a further Act of Parliament enabling them to raise more money. Towards the end of 1840, the *Journal* suggested that the slow progress of the work might almost be welcomed "as putting off the evil day when this ill-considered and rashly undertaken plan shall be called completed".

In fact it was more than two years before the work was finished. The lock gates were operated for the first time on 17 January 1842,

Previous pages:
A plan illustrating Henry Palmer's proposal for the Wet Dock.
Ipswich Port Authority

when high water was just before four o'clock in the afternoon. Palmer was on hand to superintend the closing of the gates. The first vessel to use the lock was the little sloop *Director*, outward bound for Rochester with grain and goods, which left the Dock with the mayor, John Cobbold, on board. He was, of course, also a leading Dock Commissioner. The first vessel to pass into the dock through the lock was the collier brig *Zephyr*, which had brought a cargo of coal from Newcastle-upon-Tyne. Captain Scrivenor had neglected to hoist the anchor on deck, since he expected to have to anchor in the New Cut; and as he came into the lock it caught the lock wall, to the detriment of the vessel's paintwork. "Thus", observed the *Ipswich Journal* tartly, "after no inconsiderable bungling and delay, the town may be said to enjoy the advantages of a Wet Dock".

The *Journal's* views of the event were undoubtedly coloured by its political leanings. Compare the "bungling and delay" with the comment in the *Ipswich Express* that two or three hundred people were present to witness the first operation of the gates, in spite of the dullness of the weather; and that not just the *Director* and the *Zephyr* but a number of other vessels passed through the lock "without any difficulty or accident having occurred".

The editor of the *Ipswich Journal* gleefully reported an apparent leak in the lock gates two months later, but he had to follow up with

A schooner waiting to lock out of the dock, seen in a photograph taken by Richard Dykes Alexander, chairman of the Ipswich Dock Commission, about 1860.
Suffolk Record Office

an apology, as it turned out that the "leakage" had been caused by a very high tide.

Between 1820 and 1834, the average navigation dues received by the River Commissioners had been £2,360 per annum. In 1844, one year after the opening of the dock, they doubled, while by 1855 the dues had increased to £10,000 a year. During the early years of the dock, receipts far exceeded expenditure; and the balance was used

Shipping in the dock in 1858, seen in a photograph by Robert Burrows.
Suffolk Record Office

for financing further improvements to the river, making it possible for larger vessels to reach the dock.

The final cost of the dock had escalated far beyond the original estimate. Quite how much it had all cost is difficult to determine, but in 1860 George Hurwood told the Institution of Civil Engineers that the total cost had been £130,000; when Joseph Chamberlain, President of the Board of Trade, opened the new lock in 1881, he was told in a briefing by the Dock Commissioners before the ceremony that the figure was £110,000. Whatever the final sum, it had been a highly expensive exercise.

The Stowmarket Navigation 4

THE EIGHTEENTH century saw a sudden upsurge of interest in the construction of river navigations and canals throughout Great Britain. As early as 1710, the towns of Cambridge, Thetford and Bury St Edmunds could all be reached by navigable waterways from the Wash by way of the port of King's Lynn; and within a few years Sudbury was linked with the port of Manningtree by a navigation that made use of the River Stour.

Considerable interest was shown in making improvements to rivers in various parts of Britain in order to take raw materials such as coal to the burgeoning industrial centres; and the success of the Duke of Bridgewater's pioneering work in canal-building in the Manchester area led late in the century to such an escalation in waterway construction that the 1790s became known as the decade of the Canal Mania.

Ipswich, however, remained the head of navigation on the Orwell. The town of Stowmarket on the River Gipping, midway between Ipswich and Bury St Edmunds, had no effective access to the sea, although it is known that the Gipping had been navigable in earlier times; in the 17th century, flat-bottomed barges were used to carry the bells of Stowmarket parish church down to Ipswich to be recast.

Ever mindful of the interests of its merchants, Ipswich Corporation successfully resisted proposals put forward in 1719 to make the river navigable up to Stowmarket. The well-to-do merchants of Ipswich feared that, if the Gipping was opened up, much of the trade of the port would pass the town by on its way up to Stowmarket; they controlled the Corporation, and it stubbornly continued to resist the improvement of the river.

By the 1790s early canals in other parts of the country were proving profitable, and the inland waterway network already linked London with Birmingham and with Bristol. The administration of the younger Pitt had restored the economy following the disastrous American war, and there was a new feeling of confidence beginning to make itself evident. In 1789, Robert Ransome arrived in Ipswich and set up an iron foundry that was to lay the foundation of one of the town's major industries.

Improved communications, both by land and by water, were felt to be vital if the town and its hinterland were to benefit from the new era that was seen to be dawning. Ipswich Corporation, perhaps unable to maintain its resistance any longer, decided not to contest a Bill promoted in Parliament by a number of Suffolk landowners that would authorise the making of a navigation from Stowupland Bridge on the east of Stowmarket to Handford Bridge in Ipswich.

Owners of land fronting the river promised their support, doubtless aware of the potential benefit that could accrue to them of having a navigable waterway passing through their property. The Act "for making and maintaining a navigable Communication between Stowmarket and Ipswich, in the County of Suffolk" received the Royal Assent on 1 April 1790.

Work began in May that same year but progressed only slowly, with completion targets being repeatedly put back. The Trustees appointed under the Act to oversee the construction and operation of the navigation found themselves faced with a succession of legal obstacles. These included problems arising because contractors had apparently trespassed on land over which the Trustees had no control. Further difficulties arose when it became apparent that the original survey undertaken in 1789 was somewhat inaccurate, and as a result not all the necessary work had been completed when the navigation was eventually opened on 14 September 1793.

The navigation had cost over £35,000 to construct. The total length was just over 16 miles and there were 15 locks, each measuring 60 feet by 14 feet. All the money required had been raised by loans from local landed gentry and other shareholders. Initially, four barges were employed on the river; this number was increased to ten in 1794. The tonnage charge was a penny per ton per mile on grain and other goods and a ha'penny per ton mile on coal.

The Gipping, which through a number of tributaries drains a significant area of mid-Suffolk, has always been prone to flooding. Hardly had a second Act been obtained "for effectively carrying into Execution" the original Act, by allowing the Trustees to raise further loans, than the banks and the towpath were considerably damaged by flooding, necessitating the expenditure on repairs of £2,000—more than the entire revenue for the preceding 12 months.

Trade on the navigation continued quite profitably during the first half of the 19th century, the cargoes being mainly agricultural produce taken downstream and coals being brought upriver. It proved to be quite a thriving concern, meeting the needs of the town of Stowmarket and the surrounding villages. There was even a suggestion for a canal linking Stowmarket with Bury St Edmunds, but this came to nothing.

Alarm bells sounded in 1836 when the Trustees read of proposals to build an enclosed dock at Ipswich. They made strong representations to the Ipswich River Commissioners, asking them to "admit the Stowmarket Trade into the Dock at the Bridge end . . . . otherwise a serious injury will be sustained by those whose interests we are bound to protect". They realised that the positioning of the proposed lock would mean that vessels from Stowmarket would be unable to enter the dock directly from the river. They would first have to proceed downriver past the lock entrance, before turning round to sail back upstream into the dock.

One of the Stowmarket Navigation locks, photographed by Robert Burrows.
Suffolk Record Office

What they wanted was a second small lock at the north end of the dock to provide a short-cut for barges coming down from Stowmarket, or leaving the dock to proceed up the navigation. They even offered, in an eleventh-hour letter to the Dock Commissioners in June 1841, to contribute a moderate sum "to meet the expenses attending its accomplishment". Their entreaties fell on deaf ears, and the new dock was completed the following year with a single entrance that faced downstream.

Bargeowners who attended a meeting of the Trustees of the navigation in February 1842 complained that the new lock was regularly causing them to lose a tide. Whereas they had formerly been able to reach the quays at Ipswich, discharge their cargo and return on the same tide, they were now often forced to wait until the next tide before they could return to Stowmarket. Even when they were able to leave the dock on the same tide, they could not get up to the first river lock at Handford Bridge because of the fast-ebbing tide. They also complained that the still water of the dock was sometimes frozen over before ice blocked their canal—perhaps a sign that the climate was more severe 150 years ago.

Problems such as these paled into insignificance when, in 1845, an Act was obtained for the building of a railway between Ipswich and Bury St Edmunds by way of Stowmarket. The Eastern Union Railway, which had been opened from Colchester to Ipswich earlier that year, operated the Bury St Edmunds line on its completion and was naturally keen to monopolise what traffic there was. Experience elsewhere in the country had already demonstrated that canals suffered severely in competition with railways.

Even before the line to Bury opened, the Trustees had reached agreement for the lease of the navigation to the railway company. The lease was to be for 42 years, at an annual rent of four per cent of the capital of the navigation for the first 21 years and three per cent for the second half of the period. The Trustees obtained an Act of Parliament on 26 June 1846 sanctioning the leasing of the navigation.

The Trustees were undoubtedly justified in their fears for the future of the navigation. The total tonnage of goods carried on the Gipping rapidly fell by three-quarters following the opening of the railway, although there was still a role for the navigation to play. In 1868 the waterway won back some traffic from the railway when Packards, fertiliser manufacturers with a works at Bramford and other premises in Ipswich, gave up sending their goods by rail; instead they operated steam barges on the Gipping between Ipswich and Bramford, where a small dock was constructed within the curtilage of their works.

Under the agreement between the Trustees and the railway company, the latter was required to maintain the navigation in good condition. However, the company was often in dire straits financially and needed constant prodding over the need for repairs. In 1887, the Great Eastern Railway, successor to the Eastern Union, declined to assist any further. By then the navigation was in a run-down state, and after some negotiation between the parties, the GER offered £2,000 in full settlement of all outstanding claims, an offer which was accepted by the Trustees.

The combination of the building of the dock at Ipswich and the coming of the railway heralded the beginning of the end for the Stowmarket Navigation. Trade declined over the second half of the century, and by 1900 use of the upper end of the navigation had dwindled to the occasional barge bringing the raw materials for fertiliser manufacture to Prentices at Stowmarket, returning downriver with guncotton from the explosives works there.

Soon only Packard at Bramford was making use of the river. Barges no longer used the river beyond Bramford, and Stowmarket was left wholly dependent on the railway and on road transport. The navigation was officially closed in 1922, but the last meeting of the Trustees was not held until 16 March 1934.

At that meeting, it was revealed that just £216 remained in the coffers of the Stowmarket Navigation. That residue was split equally between East Suffolk County Council and the East Suffolk Rivers Catchment Board, to whom fell the task of ensuring that the river, no longer a navigation, was maintained as an effective channel of drainage for the area.

Some of Packard's Gipping barges seen at the company's quay in the dock in 1923. In the foreground are the Humber II *and the* Yare; *the latter is now a houseboat at Pin Mill. In the background is the barque* Bellhouse, *unloading barley from Valparaiso.* Ipswich Port Authority

Trade Developments 5 in the 19th Century

In the early part of the 19th century the East Coast shipping routes were crowded with small vessels of all kinds carrying coastwise cargoes of coal, grain, agricultural produce, timber, manure and a great many other commodities. Sailing with them were larger vessels homeward bound from the far corners of the world, making for the port of London or for Hull with exotic cargoes such as tea, spices, or copper ore.

Many of those vessels, large and small, had been built in the shipyards of Ipswich, which had made a name for themselves over the years as producers of fine, fast-sailing vessels for world-wide trading. When the New Cut was made to carry the water of the Gipping past the new Wet Dock, it obliterated the St Peter's shipyard where George Bayley had built the first steamers to be launched in the town; and the construction of the dock also meant the removal of the St Clement's yard, where William Bayley was building fast little vessels for the fruit trade.

Those were well-constructed vessels in which their owners took great pride, but not all the ships that plied their trade along the coast were of such a kind. Some were so cranky and ill-found as to earn the description "crazy nail-boxes" from a correspondent in an East Anglian newspaper, who signed himself "A Suffolk Shipwright". Writing of some coasters as "nothing more than a bundle of boards", he said, "there are hundreds of vessels . . . no more fit to go to sea in their usual trade than they would be to go round Cape Horn".

It is, perhaps, not surprising that the casualty rate was high. Every storm brought its toll of ships and men, and an official report on shipwrecks in 1836 contains hair-raising stories of rotten ships and lurid tales of drunken officers, incompetent seamen, over-insurance and deliberate loss.

Casualties were not, however, confined to those with such shortcomings. Sound vessels with efficient crews could at times succumb in tempestuous weather. A sudden storm that struck in the early days of 1854 caused the loss of the barge *Emma*, whose skipper came from Chelmondiston, as she came up-Channel with stone from Guernsey. The same storm took the *James and Ann*, owned by Cobbold, Bayley and Sheppard of Ipswich, as she approached the Tyne with a cargo of grain for Newcastle.

In mid-century Ransomes were developing their trade in railway materials; and when the brig *Susan Bayley* was launched by William Bayley from the new St Clement's yard outside the dock in 1853, it

Opposite page: *The cowl of a malt kiln and the tower of St. Mary Key Church peer over the rooftops in this painting by Jabez Hare entitled St. Catherine's Quay, Ipswich.* Ipswich Borough Museums and Galleries

The Custom House and shipping at the quay photographed by Robert Burrows in 1858. Suffolk Record Office

was reported that her first voyage would be to Genoa with railway chairs presumably made in Ransomes' foundry. Most likely she would later sail from Genoa to Smyrna, or one of the other Mediterranean fruit ports, to load a cargo of dates or olives.

Raw materials, too, came into the port for the use of Ransomes and other industrial concerns. In 1846 no fewer than 90 cargoes of pig-iron, some 14,000 tons, was imported, as well as 10,000 quarters of linseed from the Black Sea and the Baltic for the dockside oil mills. Timber was also imported in large quantities.

As important as pig-iron to the expanding industries of Ipswich was the coal that fired their boilers and furnaces, and this came mainly by sea until the railways provided a more convenient route from inland mines. Ransomes had built a new foundry, Orwell Works, in the late 1830s beside that part of the river that was to become the dock, and the gasworks had already been built on a riverside site in 1822.

The coal trade had always been important to Ipswich. In the 17th century the town had a large fleet of collier cats, as Daniel Defoe

One of the stylish warehouses designed by Henry Palmer can be seen in this photograph of shipping in the west end of the dock by Robert Burrows.
Suffolk Record Office

testified. Most of the vessels engaged in the coasting and foreign trades in the 19th century were quite small, and few of the colliers carried more than a few hundred tons. Indeed, of the 25,000 vessels on the British register in 1835, only 100 were of more than 1,000 tons.

A total of 47 vessels arrived in Ipswich with coal in one week in 1847, bringing in just over 5,000 tons between them. When the barque *Ruth*, of Ipswich, brought 450 tons the following year, her arrival excited comment in the local press because of the unusual size of her cargo.

Once the fertiliser industry had become established in the Ipswich area, the first such factory having been built on the dockside about 1850, phosphates from North Africa and elsewhere provided frequent inward cargoes for the dock. Many such cargoes were sent onward up the Gipping to Packard's works at Bramford. The industry had been founded on the quarrying of coprolites, fossilised phosphatic nodules, in the Sandlings area of East Suffolk; and this material provided many outward cargoes.

In spite of all the risks they ran, some Ipswich vessels survived to ripe age, and so, too, did their seamen. In 1860 it was recorded that John Cobbold's sloop *Endeavour* was still at work as a septuagenarian — she had been built at Faversham in 1783. Her master, John Thorne, was 79 and had been at sea continuously for 67 years, and both the mate and the third hand were 65.

Life was very hard in these ships, for the crews were small and they were worked hard. The dietary scale generally adopted by Ipswich vessels in the foreign trade, according to the Ipswich Maritime Register and Almanack for 1865, was a pound of bread a day, a pound and a half of beef four days a week, with a pound and a quarter of pork on the other three days, a half-pound of flour every other day, a third of a pint of peas on pork days, an eighth of an ounce of tea a day, half an ounce of coffee, two ounces of sugar and three quarts of water a day, or "equivalent substitutes at Master's option".

Although a number of Ipswich-built ships were trading world-wide, the great majority of the vessels actually unloading in the port before the dock was built were engaged in the coasting trade. The occasional brig or schooner brought a cargo from a nearby continental port, such as the one which arrived from

Coal from the Tyne, timber from the Baltic, pigiron from Wales and all kinds of cargo came into the dock. Some of the vessels that brought such cargoes can be seen in this photograph taken in the dock by Robert Burrows in 1858. Suffolk Record Office

The variety of shipping using the dock in the mid-19th century can be seen from this photograph, another from the Robert Burrows album of 1858.
Suffolk Record Office

Gravelines in Northern France with a cargo of apples in March 1840, or the two which arrived with Dutch cheese from Medemblik and Rotterdam a few weeks before Christmas that same year.

It was generally believed that the building of the dock would lead to an increase in overseas trade, for the possibilities had been spelt out quite clearly by advocates of the dock right from the beginning of the century. "The natural advantages which Ipswich possesses over any other port on the eastern coast of Britain is mainly owing to the superior accommodation which Harwich harbour presents for the access of vessels of less than 400 tons burthen, under all circumstances of wind and weather. It is no uncommon occurrence for from 100 to 200 sail of the Baltic fleet to lie wind-bound there, 2 or 3 weeks unable to reach the several destinations, Hull, London or Liverpool, whilst with proper dock accommodation in the Orwell they might have delivered their cargoes and sailed again without loss of time," said J. Hare in a pamphlet published in 1836, recommending the damming of the Orwell at Downham Reach.

Between 1843 and 1845 the number of vessels arriving from and sailing to foreign ports more than doubled, and before the end of that decade it had doubled again. Yet even in the 1850s, when the average number of coasting arrivals was some 1,500 vessels a year, the number of arrivals from foreign ports amounted to less than 120 or 130 a year.

Of growing significance, perhaps, was the increase in size of vessels using the port, and the coming of the steamships. In the 1840s, the steamers were confined to those carrying passengers and light goods between Ipswich and London. By the 1870s steam was invading the general cargo fleet, and the Dock Commissioners were having to take into consideration the needs of an entirely new class of ship.

The following table shows how both total tonnage and the number of steamships increased:

Year	All vessels under 300T	Sailing vessels over 300T	Steam vessels over 300T	Total Tonnage
1864	1,967	9	0	118,320
1875	1,914	27	11	139,176
1879	2,218	40	20	171,385

The trend for tonnage to increase continued and a vessel of nearly 1,000 tons docked at Ipswich in 1875. The lock that had been opened in 1842 was no longer able to handle the bigger vessels coming up the Orwell, and it was becoming obvious to many that changes had to be made if Ipswich was to maintain its position as a major port.

Robert Burrows was not interested only in maritime scenes, but he took many photographs in the dock area and along the river. In this one he has caught more shipping in the dock.
Suffolk Record Office

The New Lock 6

HENRY PALMER'S proposals of 1836 included a recommendation that the lock should be sited in a sheltered spot, otherwise "in rough weather [it] would be difficult, or even dangerous, to enter." It was for that reason that the original lock was placed in the New Cut.

What he failed to consider was that the angle of the lock, as built, was far too acute for large vessels to enter the dock without difficulty. Vessels were unable to make the sharp turn, especially when the tide was flowing fast, and all too often they struck the pierheads and displaced the coping stones. On several occasions the outer gates themselves were hit.

To give Palmer due credit, however, his original plan of 1836 had shown the entrance lock sited at a more shallow angle to the proposed new channel. At some stage before construction, the angle of the lock with the New Cut was increased from about 30° to more like 45°. What influenced that change? Possibly it was the local merchants, whose own small craft were easily able to make the turn into the lock. They may not have wanted to see larger vessels plying to Ipswich and taking trade away from their own ships.

The difficulty was made a great deal worse as the size of ships increased in the course of the 19th century. By 1875, trade was beginning to fall off, and the total income from tonnage and ballast dues fell regularly year by year from £4,308 in 1869 to just £3,184 in 1874. As public anger was directed at the Dock Commissioners, the local newspapers, the *East Anglian Daily Times*—which came on the scene in 1874—and the well-established weekly *Suffolk Chronicle*, addressed the issue in a direct manner. The new daily paper declared that "the Wet Dock is one of the finest in the World, but the lock by which vessels enter is only 140ft in length. It is like a magnificent bottle with a neck so small that it is practically useless. When our plucky and enterprising ancestors—who shame the modern merchants—made the Dock they put the entrance in the wrong place. There is a turn so awkward, that even with a free lock, a fine vessel . . . runs constant risk of going aground." The *Suffolk Chronicle* used a similar metaphor: "What is the use of a splendid mansion without a door to enter by?"

In February 1875, the steamer *Lilian* was leaving the lock when one of her crew fell overboard. The master had to make the choice between running the ship aground and possibly killing the sailor in the water with the ship's propeller. He stopped the engine long enough to avoid the sailor, then re-started it just in time to avoid going aground. The screw missed the sailor's head by inches. As the

EADT grandly reported, "had he been killed, he might have obtained a martyr's fame by securing Ipswich a new lock entrance."

The hundred or so Dock Commissioners, many of whom were local merchants and shipowners, were obliquely referred to as "this most heterogeneous ruck" by the newspaper. One of them, a Mr Wiggin, who was elected by the town of Hadleigh as their representative on the Dock Commission, was particularly critical of his fellow commissioners, comparing them to "the sleepers in Washington Irving's famous story who woke up after a nap of 1,000 years to find that meanwhile the world had been going on, and that they were absolutely behind the times." The EADT further encouraged its readership with the statement that "we believe that by wise enterprise and judicious expenditure, you can make Ipswich the Liverpool of the Eastern coast."

The Commissioners were self-electing, or nearly so, and each member of the Commission tended to view matters from his own standpoint, paying more attention to his own private interests than to the interests of the town and port. Generally they were firmly set against any increase in dues, arguing that an increase such as that required to finance a new lock would persuade local merchants to bring their goods to the town by rail rather than by sea. Curiously,

The dock area as shown in a map dated 1872. The Promenade can be seen stretching from the lock to the lower dam.
Ipswich Port Authority

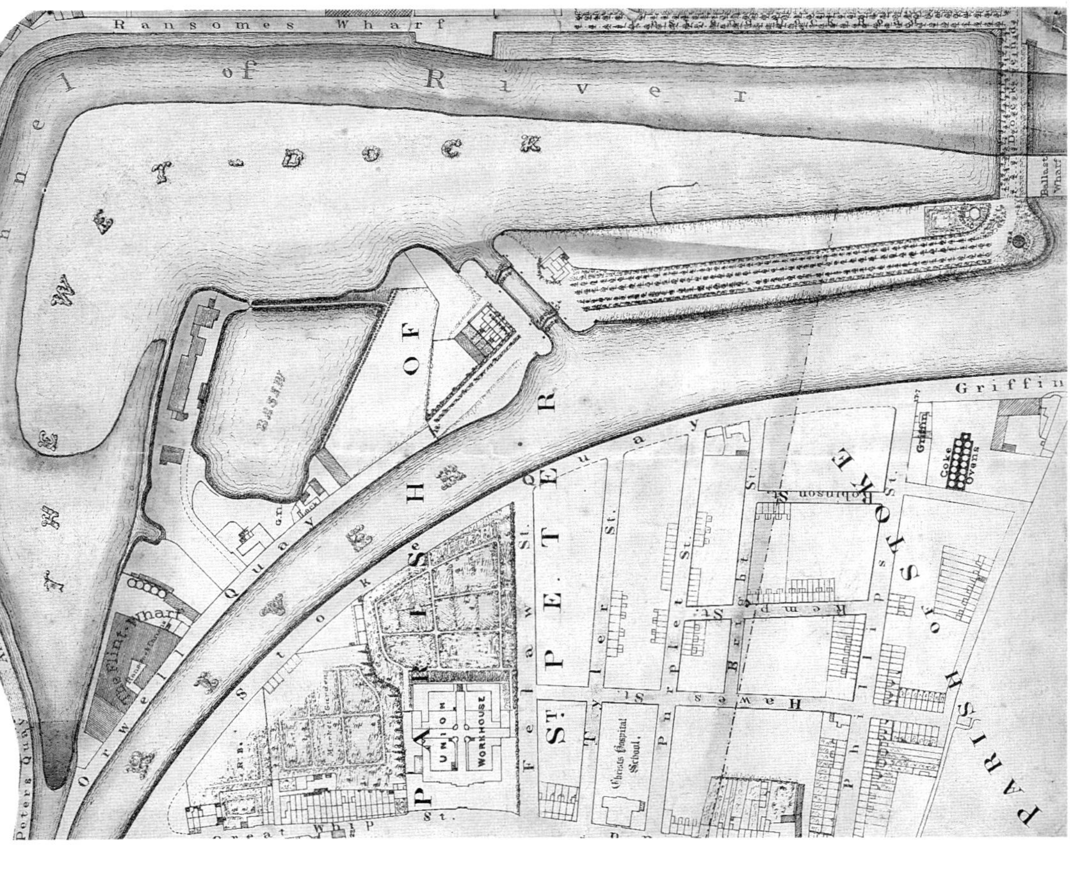

of all the cargoes handled in the port only coal attracted any duty; and in 1870 the Commissioners discovered a legal loophole by which they were empowered to charge duty not only on coal brought into the port by sea but also on that transported by rail. The 18 pence per ton levy they imposed on all coal entering the town was passed straight on to the consumer by way of higher prices. The merchants suffered not at all, and the port's coffers benefited from the additional income, though consumers in the area who had been complaining for years at the high price of coal were none too pleased. The *East Anglian Daily Times* struck a popular note with many of its readers when it spoke of the need for change at the top, suggesting that "a Bismarck is needed, whose first step would be to clear out the 100 members". Otto von Bismarck had become the first German Imperial Chancellor in 1871, and was engaged at the time in welding the German states into a united nation.

The Commissioners, many of whom were merchants and owners of small coasting vessels that were charged at a mere penny per ton to enter the port, dragged their feet. By contrast larger vessels not belonging to local owners were charged an exorbitant shilling per ton, with the object of preventing competition from these larger vessels. A special committee was set up in November 1873 to report on possible improvements, but no report had been received by February 1875.

A month later the EADT published an inspired article entitled "A Grandfather's Tale", an imaginary but highly plausible story of an elderly man talking to his grandson, presumably some time in the 1900s. Reproduced in full in the souvenir booklet produced by the newspaper on the opening of the new lock in 1881, the tale began:

> "That's where Ransomes used to be. They employed a large number of hands and did a lot of good to the town. It was a bad day for Ipswich when their works were taken away. But you see they were forced to it by the competition. With all the coal and iron they used, a waterway was essential, and the port of Ipswich was closed in 1890."

And then the grandfather was quoted as saying,

> "How do you like this beautiful piece of ornamental water? Is it not pretty in the sunlight? It's the old dock. We thought it might as well be turned to some purpose."

The fable continued in similar vein at some length, and must have helped to boost the paper's campaign for the building of a new lock. Mounting public pressure eventually persuaded the Commissioners to hold a free election for the places vacated by 24 retiring members of the Commission. The opportunity was seized upon by the campaigning newspapers to promote the names of those candidates thought to favour improvements to the port; and when the election took place on 13 August 1875, most of these candidates were successful.

Two days after the election a vessel grounded on a shoal inside the dock opposite Packard's Wharf, on the bend of the dock near

Coprolite Street. The *McLeod* had gone on the "same old spot", which had been compared in the newspapers to the celebrated Suffolk cheese, "being too big to swallow and too hard to bite, for the present miserable dredger". The new members hardly needed such an occurrence to galvanize them into action. Despite being in a minority, they took the Dock Commission by storm, and in less than a year the Committee of Management was recommending the Commissioners to apply to Parliament for powers to improve the dock.

The works agreed upon comprised the building of a new entrance lock, 300 feet long and 50 feet wide, sensibly sited at the southern end of the dock in line with the river approach. Whilst it would necessitate some dredging to provide a new channel up to the lock, this proposal seemed to be the most cost-effective way of improving the port. It was also agreed to construct a dam across the New Cut, below the old entrance lock, thus effectively extending the dock as far as Handford Lock, and making the Stowmarket navigation accessible at all states of the tide through the old dock entrance.

The Bill passed through Parliament almost unopposed in 1877, and arrangements were soon made for the necessary work to be carried out, though it was decided to defer the construction of a dam in the New Cut until the entrance lock had been constructed and the river sewage diverted. The contract for the new lock, the lock gates and machinery, and a foundation for a swing bridge over the lock, was awarded to Henry Lee & Son, of London, in the sum of £45,000. This was a firm of high repute, which had in the early

A piledriver driving the wooden piles for the outer cofferdam as work begins on the construction of the new lock, 5 November 1879.
Ipswich Port Authority

1870s been engaged on the rebuilding of Landguard Fort at the entrance to Harwich Harbour.

The first task was to build a cofferdam across the south end of the dock within which the excavation of the new lock could proceed. The contractors cleared away Thornbory's mud and gravel, and began pile-driving in June 1879. Very soon they found that it was to be no easy, straightforward job, for a deep chalk bed presented a major problem when they sought to secure a firm foundation for their work. Freshwater springs in the chalk compounded the problem, and steam pumps had to be installed to remove several thousand gallons of water a day to enable the excavation to go ahead.

Two sets of gates were installed, and provision was made for a third outward-facing set of gates to shut out abnormally high tides. The foundation for a swing bridge was provided to enable a railway to complete a circuit of the dock. Neither the third set of gates nor the swing bridge was installed at that time. The bridge was installed in 1903, but the outward-facing gates were not fitted until 1976, when they were provided as part of Anglian Water's Ipswich flood prevention scheme. The dam across the New Cut was never built, although a velocity control structure was also installed in the 1970s as part of the same scheme, to regulate the flow when there was a large amount of floodwater coming down the Gipping from the interior of Suffolk.

Under the powers granted to them under the new Act of 1877, the Commissioners built a public quay and warehouse adjacent to the old lock entrance and purchased a powerful new dredger that was used with a double shift of men to excavate the quarter-mile or so of approach channel to the new lock. The dock itself was progressively deepened to provide a depth of 20 feet in most places, though at the town end the water remained comparatively shallow.

The construction of the new lock meant that the ballast wharf had to be moved again, and stages against which ships could lie to receive their ballast were erected out from the shore inside the dock where South-west Quay was later to be constructed.

The lock was opened with great ceremony on 27 July 1881. At 11 o'clock a large invited group, including the Mayor, the President of the Board of Trade, the Member of Parliament for Ipswich and the General Manager of the Great Eastern Railway, went on board the paddle steamer *Glen Rosa*—an iron vessel operated at that time by the London Steamship Company. In her they sailed down the winding reaches of the Orwell as far as Shotley Point, the limit of Ipswich Dock Commission's jurisdiction, and then sailed back on the flood tide to reach the dock entrance at around high water. The *Glen Rosa* passed straight through the new lock, with both sets of gates open. A grandstand had been built on one side of the lock for the specially invited guests, and hundreds of other spectators lined the dock to see her enter. Cannon fired a salute, as they had done in 1839 when the foundation stone of the original lock had been laid.

Dock engineer Thomas Miller watches over the proceedings at the lock on 13 April 1880.
Ipswich Port Authority

It was indeed a great day for Ipswich. Not only was the new lock that was to herald a new era for the port officially opened, but later in the day the new Post Office on Cornhill and the Museum in High Street were also opened; and these ceremonies were followed by a grand "déjeuner" in the Council Chamber of the Town Hall, given by the Mayor for his now-exhausted entourage.

A Perambulation of the Dock 7

CONSTRUCTION OF the Wet Dock led to great changes in the appearance of the riverside area of Ipswich. With only one exception, all the buildings to be seen on the dockside today have been erected since the opening of the dock.

This clockwise perambulation of the Wet Dock starts at **Stoke Bridge**, at least the fourth bridge to be erected on this site since the original ford across the Gipping that allowed the people of Saxon Ipswich to visit their neighbours in the parish of Stoke. The route follows the line of the present quays, but occasionally digresses to record changes that have occurred since the early part of the 19th century, before the dock was built.

The **Tidemill** was a weatherboarded, timber-framed structure which stood on the town side of Stoke Bridge where the viewing platform is today. To provide the power to grind flour, it used tidal water penned in an extensive pond where the railway goods yard was later laid out. The Tidemill ceased working in the 1840s, and the weatherboarded building was moved across the road to become part of a big new steam mill, Eastern Union Mills, built by Joseph Fison, who had taken over the tidemill in 1842. In 1877 hydraulic rams were used to move the former tidemill building to another new position, when Fison built an extension to Eastern Union Mills, the machinery being supplied by E.R. & F. Turner at nearby St. Peter's Ironworks. The original building survived until further reconstruction in the early 1930s changed Eastern Union Mills into a yeast factory. The Tidemill is shown on a four-part panorama of the quays and adjoining buildings made by Edward Caley in 1837 before construction of the dock began.

The wide but shallow area of the river immediately below Stoke Bridge is known as **St Peter's Dock**, and was in regular use before the Wet Dock was opened. Its usage declined thereafter, and it is now occupied only by pleasure craft. Downstream of **Stoke Bridge** one could not reach the Common Quay along the riverside before the dock was built, for in many places the buildings came right down to the water's edge. Among these were the maltings which backed on to College Street and the adjoining St Peter's Ironworks occupied by Bond, Turner and Hurwood, the company that was to build the lock gates. This site is now occupied by confectioners **Burton, Son & Sanders**. The ironworks gave its name to **Foundry Lane**, which today runs from College Street to the quay, apparently on the line of a narrow slipway shown by Caley next to the ironworks premises.

There were many maltings between the Tidemill and the Custom House, recognisable on Caley's drawings by their kilns, topped with wooden cowls which rotated to face always out of the wind. Many of the cargoes sent coastwise from the quays consisted of malt consigned to London, Liverpool, Dublin and a host of other ports.

The upper dam was built just downstream of the iron foundry, and for more than a century the upper part of the dock has been occupied by **Cranfield Brothers**' flour mills. Much of the grain ground at these mills used to be brought to Ipswich either by sailing barge from the London Docks or by large sailing vessels which unloaded into barges in the lower end of the dock. The dock was too shallow at the upper end to accommodate the bigger ships. Subsequently, the grain was brought in lighters which were towed up the Orwell. As recently as 1975, this part of the dock was congested with as many as a hundred dumb lighters, mostly moored there side by side. It had long been the practice to use old vessels moored in the dock as floating grain stores to provide overflow warehousing. Grain tended to arrive in bulk at certain times of the year—until the late 1930s Australian wheat arrived in big four-masted sailing ships about June. The quay is now used only by

Henry Palmer's warehouse supported on cast-iron columns next to the Custom House can be seen in this photograph of the dock taken about 1880.
Robert Malster

A pair of railway shunting horses in the goods yard next to the dock. Until about 1880 horses were used to haul trains on the dock tramway.
Doug Cotton

the occasional yacht or commercially-backed converted sailing barge.

Almost everything has changed since Edward Caley made his meticulous drawings of the quays and buildings. As part of his work, he noted the condition of the quays as they existed in 1837. His comments vary from "Brick Work, piles and sheet piles in good condition" to "Brick work decayed and slightly bulged, piles decayed, no sheet piles, wattles decayed close to quay wall".

With the construction of the dock, the decrepit old quays on the town side were replaced by a new brick-faced quay, built out in advance of the old and topped by a broad roadway, which eventually reached all the way from Stoke Bridge to the site of the old ballast wharf—in what is now Helena Road. Along this road railway tracks were laid in the later 1840s to facilitate the transfer of goods from ships to railway wagons. To begin with, horses were used to move the wagons, but in the late 1880s Ipswich Corporation at last agreed to allow steam engines to cross the road by Stoke Bridge; and in this century trains were worked along the dockside lines by little tram engines which resembled self-moving goods brake vans.

As part of the dock project Henry Palmer designed and built timber-framed warehouses, which at first-floor level reached out across the quay so that goods could be loaded direct from the warehouses into vessels lying at the quay. In a report on his proposals, Palmer remarks that "it is proposed to support the front walls of the warehouses upon a Colonnade, the Columns being composed of Cast Iron". The engineer made it plain that he was anxious that these buildings should be pleasing to the eye and not merely functional, and so they were. The graceful warehouses have

Opposite page: *The Sailors' Walk, Old Custom House, Ipswich, by Frederick Brett Russel. Although Russel painted this picture in 1869, the building had been demolished in 1843 to make way for the new Custom House.*
Ipswich Borough Museums andGalleries

Opposite page: *The old Custom House that was thought to be some 500 years old at the time it was demolished to make way for a new one "more in accordance with the new character of the port".*
Ipswich Port Authority

long since given way to other, larger, buildings, but the cast-iron columns are still there on the quayside supporting new structures. These modest yet handsome warehouses were in the course of the 19th century overshadowed by lofty maltings and flour mills, which filled the space between the quay and College Street.

Near the **Custom House** on the **Common Quay** were steps which once provided access from the quay to small boats. In 1837 Caley had found the "Brick work, piles and steps very bad", and it seems obvious from the drawing that the steps were unusable. These steps were lost when the new quay in front of the Custom House was formed. The early Custom House was "a low, ill-shaped, isolated building, supported on the south, next the water, by a numerous range of pillars, reaching the whole length of the front, which is about a hundred and twenty feet, forming a colonnade, under which the masters of vessels and other seafaring people delight to perambulate", according to the historian G.R. Clarke. Writing in 1830, Clarke thought that the building was already four or five hundred years old; and when Henry Palmer made one of his early reports to the Dock Commissioners, he referred to the opportunity offered "for building a Customs House more in accordance with the new character of the port". Accordingly the building was demolished soon after the building of the dock.

In 1844-45 the Custom House was replaced by a much grander building, now the headquarters of the Ipswich Port Authority. It was designed for the Corporation of Ipswich by a young architect, John Medland Clark, who had arrived in the town five years earlier. It would seem that the scheme for building a new Custom House was promoted by Peter Bartholomew Long, clerk to the Dock Commissioners, during his second term as Mayor in 1840-41. The Ipswich Journal reported in April 1840 that the Corporation had decided to pull down the Custom House and also to demolish the crane-house and warehouses on the Common Quay, "which in consequence of the dock works were comparatively of little value in their present state", and to erect on the site "a handsome building . . . capable of affording increased accommodation as a Custom House, sufficient warehouse and other rooms for the business of wharfingers, an Excise office, and a coffee-room for captains of vessels and others connected with the shipping of the port".

The **Old Custom House**, as it is now known despite being a replacement for the former edifice, is not only an impressive building with its four columns supporting a pediment containing the borough arms, but also occupies an important position in architectural history. It is an early example of the use of red brick for a public building; earlier buildings of this nature had been built of the white bricks for which Suffolk brickyards, and in particular those at Woolpit, were famed.

Clark's plans, drawn up in 1840-41, remained with the Corporation until in 1843 the Estate Committee decided to go ahead with the building of a "Hall of Commerce" and arranged a competition for the design. It seems to have been generally accepted

THE DEVELOPMENT OF THE DOCK

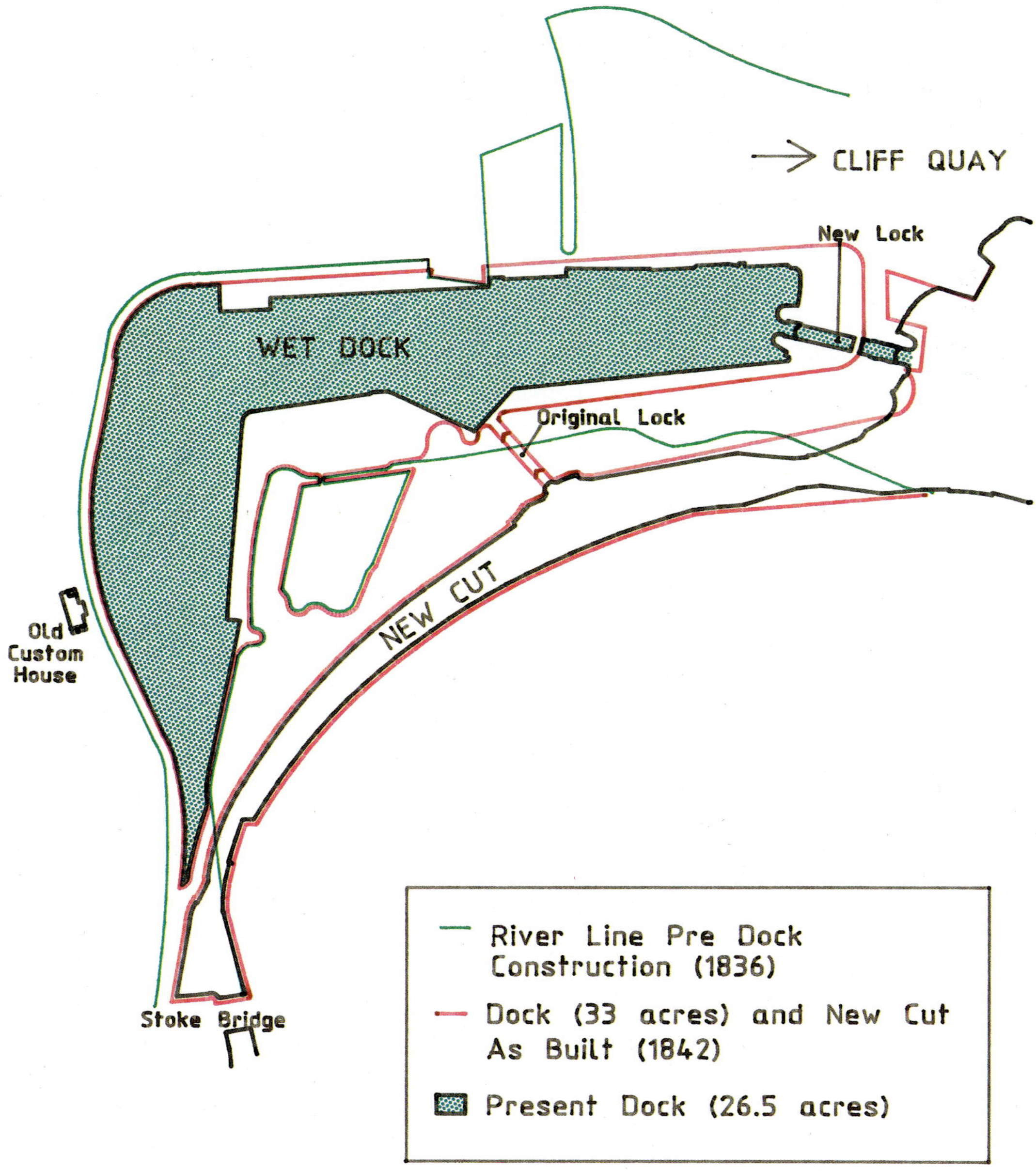

that Clark would be the winner, and so he was. The building was completed in July 1845.

On the east side of the Common Quay, R & W Paul built the towering **Home Warehouse** in the 1880s to store the home-produced barley used in their floor maltings. Barley imported from the Black Sea and later from North America was stored elsewhere.

Just beyond the Home Warehouse, which is now converted into offices for **Contship (UK) Ltd.**, are brick sheds that were built in the second half of the 19th century for Frank Alexander Christie, coal and salt merchant of Salthouse Street. In this century they were used for the storage of timber by William Brown & Co (Ipswich) Ltd. Plans in the Engineers Department of Ipswich Port Authority show how the foundations for these sheds were carried over and around the dock sewer that had been put in during the construction of the dock to carry away the town's drainage.

Before the construction of the dock, buildings completely blocked the way along the waterside beyond the Common Quay, but a lane from Salthouse Street provided access to another quay lined with brick-and-weatherboard warehouses. Here was the **Wherry Inn** that took its name from the sailing craft that in earlier days had carried passengers as well as light goods between Ipswich and Harwich; and not, despite a painting by a local artist showing a Broads craft alongside the quay, from the same type of wherry as was used in east Norfolk and north Suffolk. The Wherry Inn has gone to make way for **Isaac Lord's** coal yard, and so have some of the maltings that existed here; but the name of the Wherry has been passed on to **Wherry Quay** and to the adjacent lane. Behind the quay, around Isaac Lord's courtyard, fine Elizabethan buildings still remain.

The only buildings currently fronting on to the dock to have survived from pre-dock days are the little malt kiln, preserved today as the **Malt Kiln public house,** and the adjoining malting in which the Wet Dock 150 Exhibition is being held during 1992. William Horn, sailmaker, had his sail loft near here in a black weatherboarded building that also housed the Anti Dry Rot Company, which in the 1840s urged the Dock Commissioners to allow them to treat their new lock gates.

Beyond Wherry Quay is the area now known as **Neptune Quay**. Here were the premises of **James Whitmore**, ship chandler and sailmaker, where sails were still being made for spritsail barges into the 1980s. Whitmore's business had transferred from Harwich at the end of the 19th century. This tall building is now occupied by a snooker club and dockside bar. On the other side of the lane linking Neptune Quay with Fore Street was the coalyard of **Mellonie & Goulder**, which had a gantry by which coal unloaded from ships at the quayside could be taken into the yard.

Abutting the coalyard used to be a large malting belonging to Meux's Brewery, of Nine Elms, London. This replaced earlier, smaller buildings on this site in the 1920s; among the premises which disappeared with the building of the maltings was the **Pilot**

Opposite page: *A plan of the dock showing the line of the river before the dock was built, the configuration of the dock in 1842 and the line of the present-day quays.*
Ipswich Port Authority

A motor coaster lying alongside Eastern Counties Farmers' silo. At right is the first artificial fertiliser factory built in Ipswich in the mid-19th century, now demolished.
Robert Malster

Inn. When these maltings were demolished in the late 1960s, the site was taken over by Hall & Co for use as a ballast works. Sand and gravel dredged from the River Stour was brought to the quay by lighters and discharged by a floating crane.

An adjacent site was occupied up to the 1930s by the Steam Packet Tap and a variety of commercial premises. These were taken over by the co-operative, **Eastern Counties Farmers Ltd,** who traded from here until the late 1980s. Today this area, along with that of Mellonie and Goulder's former coalyard, is awaiting development for amenity housing or offices.

At the time the dock was built, William Bayley occupied **St Clement's Shipyard**, between the river and Duke Street. The building of the new quays as part of the Wet Dock scheme sealed off the inlets housing the slips of the shipyard. One of these was a patent slip or marine railway on which vessels could be hauled out on a wheeled carriage. The building slip was on the same line as the entrance to a later fertiliser works yard, set at an angle to the quayside roadway. With the construction of the dock, the shipyard was moved to a new site on reclaimed land just below the lower dam and outside the dock.

On the site of the shipyard Edward Packard set up the first plant in the Ipswich area for the manufacture of artificial fertiliser. In the mid-19th century a new road was driven through from the quay to Duke Street and named **Coprolite Street**, from the raw material used for the fertiliser industry.

The Ipswich factory did not remain in operation for many years, for it seems that the stench created when the ground coprolite was mixed with sulphuric acid on a brick floor proved too much for the townspeople; and in 1854 Packard built a new works for the manufacture of superphosphate at Bramford, the Ipswich premises being retained as a store. Edge-runner stones that were used to grind the coprolite in Packard's works are to be seen today at the rear of Suffolk College.

Between the site of St Clement's Shipyard and yet another inlet was the building in which William Worby set up a branch works for J.R. & A. Ransome. The name "Orwell Iron Works" can be seen on Caley's drawing, painted on the end wall facing the river. Also visible on the drawing is an iron crane, perhaps made by Ransomes, standing on the adjacent quay. Later the whole area between the dock and Duke Street was to be occupied by **Orwell Works**, to which Ransomes transferred all their business in 1849. The completion of the move was celebrated with a great feast in one of the new buildings, which displayed the motto, "Success to the Plough and Rail", appropriate for a firm which had made its name with ploughs and had then diversified into railway engineering. In subsequent years much of the firm's raw materials came in by sea, and many ploughs and thrashing machines were loaded on to rail wagons on the dockside tramway for dispatch to the London Docks.

It was only after the Second World War that Ransomes, Sims & Jefferies moved to a new works at Nacton. Many of the old buildings of Orwell Works still survive, some of them used by the Port Authority and other dock users for warehousing purposes.

Alongside is a new section of quay known as **Orwell Quay**, built in 1968 for a weekly roll on/roll off service to Hamburg. Whilst no longer used for that purpose, the quay still plays an important part in the Port Authority's operations.

Beyond Orwell Quay, there was at the time the dock was built a deep inlet with the gasworks at its head. The gasworks of the Ipswich Gas Light Company had been built on this site in 1822, with the idea that the coal needed for the production of gas should be brought in by sea. The Gas Dock, as it was known, survived for some years after the construction of the dock, but was eventually filled in to form a site for an extension of the gasworks. In the early days the coal was unloaded by hand, a somewhat time-consuming process;

but in the 1880s two hydraulic cranes were installed on the gasworks wharf to unload the coal by grab; hydraulic pressure was supplied by a plant in the gasworks itself.

The continuation of the quay from the gasworks wharf as far as **Ship Launch Road** is known as **Helena Road**, after the former brig-sloop *Helena* which in 1868 was moored in this part of the dock to serve as a floating church. The ship church remained at Ipswich until 1880. During the period the *Helena* was moored there, Helena Road was a tree-lined thoroughfare from which access could be gained across the lower dam to the **Promenade** at the eastern end of the newly built island.

Construction of the new lock in 1880-81 did not affect the popularity of the Promenade, which consisted of pathways between three rows of poplars and limetrees planted out in 1846. It was a much-appreciated amenity, although there have always been those who have scant respect for what is regarded as public property. In the 1840s, the Dock Commissioners offered a reward of five guineas for information regarding "some evil disposed person or persons" who had maliciously damaged and broken "one of the Iron Seats placed for the accommodation of the Public at the Walks of the Promenade near the Lock Gates".

At the south end of the Promenade was the **Umbrella**, a canopied shelter, and a statue of Pegasus, the winged horse. Close by was an

The octagonal cottage in The Avenue, close to where the new lock was constructed in 1879-81, photographed by Robert Burrows in 1858.
Suffolk Record Office

octagonal cottage, the residence of the "curator" of the Promenade, which can be seen in photographs by Richard Dykes Alexander, an early amateur photographer and chairman of the Dock Commissioners. The cottage was pulled down in 1904 to make way for new railway tracks, and eventually the Promenade itself gave way to railway sidings during the latter part of the 1920s.

There were originally no quays on the south and west sides of the dock. Just after the turn of the 20th century, over a thousand feet of new quay was added to cope with the increased tonnage wishing to come to the port. **South-west Quay** linked the former lock with the new one, replacing old "dolphins", wooden berthing stages, that had been installed soon after the dock had opened.

The site of the old lock can still be seen, for although a quay, **Timber Quay**, was built across the inner end of the dock as part of this development, the knuckles of the entrance from the New Cut survive. Beside the lock were built the **Harbour Master's house** and the **Lock Cottages**, to add point to the opinion expressed by the Commissioners in 1842 that the men in charge of the lock should "be present at all times". Both still stand, occupied by Port Authority staff, including the present Harbour Master.

Further along the New Cut roadway was the **Lock Tavern**, which began life as Westgate's Commercial Hotel at a time when the Promenade was a fashionable resort of the townspeople. The Lock Tavern closed in September 1955, and was later converted into flats for Dock Commission staff. The building was eventually demolished in 1984.

There were a number of industrial premises on the neck of land formed between the upper and lower dams of the dock, known nowadays as the **Island Site**. These included George Tovell's roman cement works, which seems to have been operated by water power. The mill pond in which tidal water was impounded to work the mill survived the building of the dock, a new channel being made from the New Cut to take water through sluices into the pond. Later, when much of the "island" between the dock and the New Cut was given over to timber importing, the pond was used to store large baulks. Other similar timber was stored afloat on the Wherstead side of the river. In this century, the bank separating the pond from the dock was removed, and the former mill pond then became known as the branch dock; but it was eventually filled in as the quays were extended.

Tovell's Wharf no doubt took its name from George Tovell, owner of the cement works and also a Dock Commissioner. To the west, at **Flint Wharf,** Frederick Ransome had in the mid-19th century an artificial stone manufactory in which he produced materials for the building trade. His products no doubt made use of the output from Mr Tovell's mill, and in the 1880s Ransome took out patents for a rotary cement kiln of his own invention.

Thereafter the Dock narrows to a point, defining where the original river that used to wind through the town was dammed in constructing the new Wet Dock a hundred and fifty years ago.

COAL WHARF
A. HUBERT
WHARF

Dock Inns and Beerhouses 8

A HUNDRED years ago, every maritime town in the country could boast of numerous inns, taverns and public houses clustered around its harbour or docks, and Ipswich was certainly no exception. Maps and directories of the 1890s show more than 50 hostelries within 200 metres of the quays. Several of them, indeed, were to be found among the warehouses, maltings and mills that lined the quays themselves.

Only a few of them remain today, notable examples being the **Ship Launch** in the road named after it, the **Lord Nelson** in Fore Street and the **Old Bell** on the Stoke side of Stoke Bridge. Many once well-known hostelries, such as the **Crown** in Greyfriars Road and the **Angel** and the **Lion and Lamb** in Fore Street, have disappeared under road-widening schemes. Others have simply given up the licensed trade, and are now used for other purposes.

The **Angel**, which stood on the corner of Angel Lane and St. Clement's Church Lane facing the **Lion and Lamb**, had a very long history as a hostelry. In pre-Reformation days the **Angel** was the principal inn of the neighbourhood, and it is reputed to have been a meeting place for Cistercian monks. It closed in 1900.

The Lion and Lamb and, on the other side of Angel Lane, the Angel, from a photograph by William Vick.

Towards the end of the 19th century there were no fewer than five public houses on the short stretch of dockside between Albion Wharf and Coprolite Street, serving the needs of dockworkers and seafarers. To the west of the Custom House were the **Dolphin** and the **Union Jack,** and to the east of it were the **Wherry**, the **Pilot** and the **Steampacket Tap**. The latter, on the corner of Coprolite Street, was an adjunct of the **Steampacket Hotel** in Duke Street.

Key Street—or Quay Street, as it was formerly known—and Salthouse Street, running just behind the dockside warehouses and maltings on the town side, had the greatest density of pubs. Four of them, the **Bull**, the **Gun**, the **Maltster's Arms** and the **Ram**, were no more than a few dozen metres apart. The **Bull** dated back to the 16th century, and ceased to be a pub in the 1960s. In 1916, a Zeppelin's bomb destroyed part of the roof of the **Bull** as well as an adjoining cottage.

The **Maltster's Arms** had a long courtyard and garden which, before the building of the dock, led down to the water's edge. The **Ram** was traditionally said to have been the birthplace of Henry Tooley, the Ipswich merchant whose considerable fortune on his death in 1551 provided for the building of almshouses in Foundation Street, whose Victorian replacements can still be seen today.

Opposite page:
The Pilot Inn, one of several public houses along the quays serving the needs of dockworkers and seamen, seen in a photograph taken in the 1890s.

Opposite page: *Isaac Lord's Granary, by Alan Waddington Bellis. The granary and the other buildings in the group lying behind 80 Fore Street survive from the days when merchants such as Henry Tooley became rich on imports of Gascon wine and Icelandic cod.* Ipswich Borough Museums and Galleries

Opposite page, lower picture: *New Cut West and the entrance to the Old Dock, by John Moore. A vessel is seen entering the original lock, on the right, replaced by the new lock in 1881 but not immediately filled in.* Ipswich Borough Museums and Galleries

Pages 58-59: *The extent of the dock can be seen in this 19th-century lithograph. The gardens in the foreground belonged to the old Union, the workhouse whose boundary fences were affected by the making of the New Cut.* Ipswich Borough Museums and Galleries

The landlord of the **Ram**, Thomas Nicholls, was killed on 12 December 1778, long before the dock was conceived, in a fight between the Press Gang and a number of men from the nearby **Green Man Inn**.

When the River Commissioners were formed in 1805, they found a convenient meeting place for their smaller committees in the **Green Man**, whose landlord was paid the then-generous sum of five shillings each time this room was used.

Another of the dock area pubs, the **Anchor**, stood on what is now the site of Kennings Car Hire in Duke Street. In 1836 the owner of the brig *Maria* complained to the Commissioners that the pilot, Robert Welham, "having seen the vessel ashore and left her, failed to put in an appearance on the following tide". He was eventually discovered in the **Anchor** with the crew of the vessel, "singing a song". It was already shortly after high water, and, by the time he and the ship's company had struggled over to the quay, the tide had fallen a foot. "I trust such gross neglect will not pass unnoticed", said the *Maria's* owner in a stern letter to the Commissioners. It did, in fact, because the subject was crowded out of the next meeting by lengthy discussions on a new Dock Bill.

Clearly, the temptations of the inns and taverns around the dock were too much for some folk. Back in 1807, the first engineer employed to maintain the River Commissioners' new steam dredger had been "lately guilty of great irregularities, particularly being in a state of intoxication when he ought to have been assisting in the repair of the engine".

One of the most famous of the dock area inns was the **Neptune,** in Fore Street, that has given its name to Neptune Quay. Originally the residence of an important merchant, it dates back in part to the 15th century, and was brought up to date in 1639, a date which appears on the timbering at the front of the building. Just when it became an inn is unknown, but in 1777 when the property was leased by Thomas Moore to Daniel Orford it was described as "lately called **Neptune Inn**".

Unusually for a pub, it retained much of its elegant carved oak panelling and its fine plastered ceilings. In 1888, Dr J.E. Taylor wrote that "perhaps the most notable part is the present tap-room. The quay labourers sit over their beer in a carved-ceilinged, oak-pannelled room, the mantelpiece of which is good enough to lead to the sin of covetousness". Those same labourers used to receive their pay in the bar parlour, and there can be no doubt that a good proportion of that money was paid over to the landlord. Much of the oak panelling and other carved timbering from the interior was removed in 1924, but the **Neptune** retained its licence until 1937. It was acquired in 1947 by Mr George Bodley Scott, who set out both to restore the building and to replace much of the panelling and other woodwork. The oak panelling in the ground floor, which he made himself in 1960/61, closely resembles that of the original.

AAR

20th Century Trade 9

IN THE first year of the new century it is recorded that 2,331 vessels used the Port of Ipswich. The average deadweight tonnage of vessels was still below 100 tons, a figure that was not to be exceeded until 1911. As Cliff Quay was built and extended from 1925 onwards, the average tonnage leapt upward—155 tons in 1927, 202 tons in 1932, 309 tons in 1950—and by 1959, two years after completion of the final stage, it had increased to 387 tons.

These figures reflect not only the larger size of vessels that were able to use the port, but also the increase in vessel size generally. Even if there had been no development outside the dock, average tonnages would still have increased. The dock and lock may have their limitations, but large vessels can nevertheless still be accommodated. When the tide is high enough, a vessel can pass through the lock with both sets of gates open, and one longer than the 300 feet of the lock itself can reach the quays. The *ss Sheikh* established two records in the early 1920s: at 389 feet she was the longest vessel to enter the dock; and at 2,828 tons, she was also the largest vessel to have passed through the lock. The longest vessel to berth within the dock, the 392-foot *ss Virginia*, came in 1935, and five years later the *ss Szent Gellert*, 2,915 tons, established an all-time tonnage record.

During this century, facilities in the dock have been steadily improved by the construction of South-west and Timber Quays in 1904; the creation of End Quay and Tovells Wharf in 1923/24; and the rebuilding of Public Warehouse Quay in 1965. The swing-bridge across the lock was replaced in 1949 by a new bridge built by the Ipswich-based company Ransomes and Rapier, formed in 1869 to take over the railway materials department of the original Ransomes firm; the new bridge enabled main line rail traffic to reach Cliff Quay. Three years later, anticipating the coming surge in general cargo traffic, a new warehouse was built at South-west Quay.

In 1968, following the departure of Ransomes, Sims & Jefferies from Orwell Works to new facilities on the outskirts of the town, the Dock Commission purchased most of their property between the quay and Duke Street. Former foundry buildings were modified into warehouses and a roll on/roll off berth was built. Sadly this facility had but a ten-year lifespan, owing to the closure of the service for which it was designed. Since then, the quay has been used only occasionally for timber, paper and steel imports. More recently it has been in use as a lay-by berth.

Twentieth-century trade in the dock has been based largely around industries that were represented in the immediate locality of the port. The privately-owned Gasworks Quay was able to

Opposite page: *The German schooner* Aar *in the dock in the 1970s. The last foreign sailing vessel to bring a cargo to Ipswich, she was sold shortly after her last visit to the dock to become a cruise ship in the Caribbean.*
Robert Malster

accommodate vessels of up to 1,400 tons deadweight, and as the town grew in size and population the demand for gas rose accordingly. In 1910, no less than 37,173 tons of coal was imported, a figure that was to rise steadily, with the exception of the immediate post-war periods. Imports of coal for the gasworks reached a peak of 113,362 tons in 1958 and held around 100,000 tons until the coming of natural gas to the area forced the closure of the works in 1970. For the last 15 years of the gasworks' life a small export trade in coke breeze to Scandinavia had been established, although this never exceeded an average of 500 tons per week. The Gasworks Quay fell into disuse in the 1970s, but in recent years it has been resurfaced and used by Anglo-Norden Ltd for the import of packaged timber.

South-west Quay was built in 1904 on the site of the then Ballast Wharf, and it was appropriate that Tilbury Roadstone, trading under the name of EARAT (East Anglian Roadstone and Tarmacadam), should have chosen this site for their Ipswich depot in 1925. Cargoes of stone, mainly granite, were imported from Scotland, Cornwall and the Channel Islands and converted into roadmaking materials at their coating plant on the quayside. At the peak of the trade, some 40,000 tons of granite was imported annually. The company ceased trading at this site in 1984; the depot

The mv Warwick Brook *unloading coal at the Gasworks Quay. The cranes here were operated by hydraulic power.*
Ipswich Port Authority

was taken over by Tarmac Roadstone, who still operate a coating plant there, importing between 14,000 and 20,000 tons of granite and other stone per year.

Cranfield Bros. Ltd was founded in 1884 with a mill of one ton per hour capacity, situated at the end of the western arm of the Wet Dock. The business soon expanded, and in 1906 the mill was completely rebuilt and production increased ten-fold. Five years later, when their No.2 Mill came into operation, production was again doubled to nearly 20 tons a day. These dockside premises imported large quantities of high-quality foreign wheat for many years.

During the earlier part of the century, many cargoes came from the west coast of America and from Australia's Spencer Gulf ports in large sailing ships, many of them four-masted barques. These often lay at South-west Quay and unloaded the wheat overside into sailing barges which lightered the cargoes up to the mills. The larger steamers had to lie at the buoys in Buttermans Bay, in the Orwell below Pin Mill, to unload their cargoes into the barges.

By the 1930s, large sailing ships were becoming a thing of the past, and only Captain Gustav Erikson continued to operate a fleet of square-rigged vessels at a time when others had all turned to steam. His vessels were occasional visitors to Ipswich Dock with

Australian wheat being unloaded from the barque Penang *in the dock in 1933. The sacks are being tipped down chutes into the hold of a sailing barge which will take it up to Cranfields' mill at the head of the dock.*
J.A.E. Burrows/Richard Smith

Opposite page: *The American four-masted barque* Dirigo *unloading a cargo of grain in the dock. She was the first steel four-masted barque to be built in the USA.*
Richard Smith

Right: *Loading sacks of malt into a sailing barge in the New Cut in 1929.*
Doug Cotton

Below: *Unloading timber from a sailing barge at Flint Wharf, also in 1929.*
Doug Cotton

cargoes from Spencer Gulf, the *Melbourne* coming to the port in 1932, a year before her loss in collision in the North Atlantic. When she was thrown open to the public on 21 June 1932, nearly 5,000 people took the opportunity to go aboard her.

In the 1950s cargoes for Cranfields were normally received at Cliff Quay in consignments of up to 7,000 tons, and were brought to the millside by the firm's own lighters. In more recent years, home-produced wheat has been favoured, and this is brought in by lorry.

Pauls PLC, as they are presently known, trace their origin back to 1850, though it was not until 1893 that the company started to use the port for importing barley and other grains. Many cargoes of malting barley came into the port from the Black Sea and later from North America to supply the maltings that line the dock, and malt was a frequent outward cargo. The company began importing maize for the manufacture of animal feeds in 1907, and this constituted many cargoes from that time onwards. In more recent years, the demand for compound foods has grown; and a new trend of imports emerged in the 1960s, coupled with the rapid rise in home-produced grains. Oilcake, beans, soya meal and other non-grain commodities were imported at Pauls' Albion and Eagle

The timber raft Amerker Brug I, *seen here in a watercolour by E. Owen Clark, arrived in Ipswich on 26 April 1919 from Trondheim in Norway. The experiment of importing sawn timber in this way was not repeated.*
Ipswich Borough Museums and Galleries

Mill wharves. Nowadays imports are seldom seen at these quays, but exports of malt leave from Albion Maltings on a regular monthly basis for Germany.

The farming co-operative, Eastern Counties Farmers, for many years occupied a site in the north-east corner of the dock. They established a compound feed mill on this site in 1954, and until the mill closed in the mid-1980s, a variety of cargoes were loaded and unloaded at their curving quay by Coprolite Street.

Timber has for many years been an important commodity for the port; since the First World War it has been in the hands of two companies, Gabriel, Wade and English Ltd and William Brown & Co. Ltd. Gabriel, Wade and English established their Ipswich branch in 1919, and one of their first imported cargoes was a 360 feet long raft, weighing 4,750 tons, and consisting of 200,000 cubic feet of timber, which was towed from Trondheim in Norway to be discharged in the dock. When Cliff Quay was built, the company became the Dock Commission's first tenant there, occupying a site continuously until their merger in the early 1970s with William Brown & Co Ltd.

This latter company occupied a site behind Tovells Wharf and End Quay from soon after the filling in of the Branch Dock and the

Demolition of dockside buildings paves the way for the construction of Orwell Quay.
East Anglian Daily Times

Previous pages: *A view of the Wet Dock taken in 1963 from the top of the waterless gasholder that had been erected at the gasworks in 1928; it was the first of its type to be built in Britain.* *Ipswich Port Authority*

construction of these quays in 1923. Imports grew steadily and by 1969 William Brown's were handling 65% of the port's timber, some 26,000 tons. By the late 1970s, however, imports of timber in the dock virtually ceased, the company preferring to deal with smaller East Coast ports. Their storage sheds and processing plant were removed and the vacant site began to be used for long-term container storage and container repair.

End Quay and Tovells Wharf were seldom used in the latter part of the 1970s until, in 1981, a contract was won by Forest Thinnings Ltd to supply timber to, of all places, Sweden. A rare example for the port of "sending coals to Newcastle", this trade has continued until the present day with modern self-loading vessels exporting annually some 25,000 tons of logs, grown locally in Thetford and Rendlesham Forests.

Most of the trade of the port is now carried on outside the old dock, which has undergone a metamorphosis during the past 20 years. Where sailing barges used to load and unload their cargoes of malt, wheat, tiles and timber, yachts now lie alongside the floating jetties of a modern marina; and where maltsters used to labour sweatily turning the steaming malt on the kiln floor, office workers and others now sup their ales.

Below: *An aerial view of Orwell Quay under construction in 1968. A collier is unloading at the gasworks quay.* Ipswich Port Authority

The Growth of the Modern Port 10

AS THE 19th century drew to a close, the role of the Ipswich Dock Commission was mainly that of port operator. They could hardly be called landowners, since the Commissioners were responsible only for the land between the New Cut and the dock—now known as the "Island Site"—and that area surrounding the new lock entrance.

The Ipswich Dock Act of 1898 allowed for the replacement of the then ballast wharf by a concrete quay, but this was by no means the only enterprising idea that was under consideration. In March 1901 the civil engineer C. A. Brereton proposed that an outer "holding" dock or tidal basin should be built in Cliff Reach where vessels could be berthed alongside while waiting for quays in the dock to become vacant. At that time, vessels bound upriver for Ipswich were forced to wait at anchor off Harwich, if there was insufficient quay space for them in the dock, and it was felt that some better arrangement needed to be made.

Another scheme of Brereton's was for the construction of a second wet dock, on the site of what is now the West Bank Terminal. Voting on these two proposals was very close, and, although many of the Commissioners approved of these schemes both were eventually abandoned in favour of the building of a new quay along the western side of the dock, where the ballast stages were situated. By the turn of the century there was no longer the same need for ballast, for much of the coasting trade was in the hands of flat-bottomed sailing barges which did not require ballast, while an increasing proportion of the foreign trade was carried in steamers.

This 800-foot ferro-concrete quay, completed in 1904, was named South-west Quay from its location within the dock. The construction of the new quay and the laying down of railway lines to serve it led to the filling in of the original lock entrance, which had survived at least in part into the new century.

The amount of tonnage handled and the number of ships entering the dock had remained almost constant since the opening of the new lock, an indication that saturation point had been reached. Careful decisions now had to be made as to whether the port should make the most of its existing facilities, improving them where possible, or should develop outside the dock in the tidal River Orwell.

After much discussion a further Dock Act was obtained in 1913 to enable the Commissioners to build a new lock, larger and more suitable for the big steamers that were by then using the port, on the site of the 1881 lock. The scheme, which would have enabled any ship that could enter Harwich harbour to enter the dock, was expected to cost some £300,000.

During the First World War, the dock was used by a variety of naval vessels which came for repairs and refitting by Ipswich engineering firms. Among them were submarines which lay in the dock while their engines were overhauled by men from the Diesel Works in Hadleigh Road, later owned by Vickers-Petter Ltd. At the end of the war a captured German U-boat was moored in the dock on view to the public.

The outbreak of war in 1914 prevented the implementation of the proposal to enlarge the lock, and when the time came at the end of the war for the Commissioners to go ahead they had second thoughts. The huge increase in labour charges and the cost of materials during the war would mean that the proposed dock would cost a great deal more than the Commission could hope to recoup from extra revenue. Instead of constructing a new lock that would permit larger vessels to enter the dock they decided to develop deep-water berths downstream of the dock.

During earlier discussions the suggestion had been made that development should be carried out on the western side of the river, which seemed to be the natural choice. When the time came to go ahead the eastern side was chosen.

The Commissioners were able to purchase some 50 acres of Greenwich Farm when it came on the market after the war. The farmhouse was demolished and the land used for the construction of a new quay; land behind the site of the quay was levelled to provide space for the development of port-related industries. Overnight the Commissioners had become real landowners.

The Ipswich Dock Act of 1918 sanctioned the purchase of the land, and permitted the construction of a tidal quay that would enable ships too large to enter the dock to berth alongside. The Act also gave the Dock Commission powers to build a further 1,000 feet of quay inside the dock; construction of this new quay involved the filling in of the old branch dock and so provided much-needed new land area that could be used for commercial purposes.

Construction of Cliff Quay, the new deep-water quay downstream of the dock, began in 1923, and the first section—600 feet long — was opened in 1925. There were high hopes that this tidal quay would be the salvation of the port, but the number of vessels using Ipswich actually decreased until 1932, although tonnages were higher, a reflection of the ability of Cliff Quay to accept larger vessels.

The turning-point came in 1933, as Britain began to emerge from the Depression, when two companies, Fisons and Shell-Mex, leased large areas of land behind Cliff Quay on which to build their own installations. The fertiliser and oil trades were to be the port's mainstays for the next 30 years.

In spite of the gloomy economic climate of the time it was apparent to the Dock Commission even in the 1920s that serious congestion was likely at Cliff Quay when more trade was attracted

Right: *Concrete sheet piles form a foundation for the new Tovells Wharf in April 1924. This and the other new quays were built out into the dock.*
Ipswich Port Authority

Below: *A view of the lock and the west bank of the Orwell taken from a gasholder in the 1920s. The first section of Cliff Quay can be seen on the left.*
Ipswich Port Authority

Between 1923 and 1925 considerable changes were made within the dock area, the old branch dock being reclaimed to provide more space for timber storage and new quays being constructed. In the picture on the left earth is being tipped to fill the branch dock; pre-cast concrete piles lie stacked ready for use. Below is the works office and carpenters' shop used during work on the new quays.
Ipswich Port Authority

to the port, and as soon as the first stage was completed work began on a further 600 feet of quay. This second stage was completed in 1929, and then during the early 1930s work began on a third section that was to take the quay to a total length of 1,800 feet; this work was completed in 1938.

Wartime naval links were recalled in 1935 when two minesweepers belonging to the 1st Minesweeping Flotilla at Portland, HMS *Halcyon* and HMS *Dunoon*, entered the dock on the first official visit to the port to be made by vessels of the Royal Navy. The two warships were berthed at End Quay for five days, during which their crews were entertained in the town; all the cinemas in Ipswich offered free seats to the seamen.

In 1937 Ipswich Corporation decided to build a modern coal-fired power station downstream of Cliff Quay to supply electricity to the borough's increasing number of customers. Work on the site began in 1939, but was interrupted by the outbreak of war and did not recommence until 1945. By the time the station was opened in 1949 the electricity supply industry had been nationalised, and it was under the British Electricity Authority that Cliff Quay power station commenced commercial operation. A jetty was constructed to enable 4,500-ton colliers to discharge coal for the station, which in its first year of operation consumed 244,000 tons of pulverised coal. During the ensuing 30 years colliers belonging

A view of the end of the branch dock, with one of the contractor's engines used to haul trains of tipper wagons about the site.
Ipswich Port Authority

A fire float belonging to the National Fire Service was sunk during one of the raids on the dock area. It is seen here being salvaged after the raid. David Kindred

to the BEA and its successor, the Central Electricity Generating Board, made a substantial contribution to the revenues of the port; one of these colliers was appropriately named *Cliff Quay*.

With the outbreak of the Second World War, commercial use of the dock fell away to a considerable extent, but this did not mean that the dock ceased to be a bustling place. Quite the contrary, for HMS Bunting, a base for armed patrol trawlers, was set up in the port, and the almost daily comings and goings of the converted fishing vessels enlivened the local scene. One of them, HMT *Adonis*, master ship of one of the two flotillas at Ipswich, was sunk by an E-boat off Lowestoft during the night of 14 April 1943, with the loss of 21 of her crew of 32.

Dockside premises suffered in bombing raids, one attack on 2 June 1943 setting a gasholder on fire at the gasworks. During this raid one of the attacking aircraft, a Focke-Wulf FW190, struck a crane jib and crashed. On two occasions in the 1950s, unexploded bombs were dredged up.

Opposite page: *The dock in the 1970s. The upper picture shows a Dutch coaster unloading a cargo of fertiliser at the Eastern Counties Farmers quay, with the tower of St Clement's Church visible beyond the ballast plant, and the lower picture shows the motor barge* Celtic *entering through the lock; in the background work is proceeding on the construction of the West Bank Terminal.* Robert Malster

The final stage of construction at Cliff Quay itself was also delayed until after the 1939-45 war. This fourth stage, an additional 1,450 feet of quay requiring substantial land reclamation on a bend in the river, was finished in 1957, having cost nearly £650,000. It was formally opened on 10 May 1957 by Airey Neave, then Parliamentary Secretary to the Ministry of Transport and Civil Aviation, who was later to die tragically in an IRA bomb attack in London.

The most recent development in the port has been on the Wherstead side of the river, where the West Bank Terminal has been opened on a site reclaimed by the dumping of gravel dredged from the river. The mudflats were purchased in January 1973 from

59

CELTIC

R&W PAUL LTD

SCONE
ROCHESTER

The 150th anniversary of the opening of the dock was celebrated on 17 January 1992 when the Ipswich Port Authority dredger Samuel Armstrong *carried the Mayor of Ipswich and other guests across the dock to the Old Custom House for a reception. At left the* Samuel Armstrong *approaches the Common Quay, and below is the view from the dredger of the Home Warehouse, now converted into offices for Contship (UK) Ltd.*
Robert Malster

Ipswich Council for £33,600, an amount that certain councillors believed was far too cheap; others pointed out that it was a good price "for a piece of mud". Construction started immediately and the first vessel, the 416 feet long *Ipswich Pioneer II*, arrived to load trailers on 20 November 1973. The Terminal has been extended several times since its opening and now covers 30 acres, with a roll on/roll off terminal handling 125,000 trailers each year to and from Rotterdam, and a container terminal handling 45,000 units each year to and from Zeebrugge.

Whilst the use of the Wet Dock has diminished with the advent of larger vessels and the development of the terminals downstream, it still plays an important role in the life of the port. Grain, malt, timber and stone are still handled across the quays, and there are growing leisure activities, with a marina, sheltered berths for historic craft, and also a berth for the lightship TS *Orwell*, now the training headquarters of the Ipswich Sea Cadet Corps. Other training ships such as those operated by the Sail Training Association are quite often to be seen in the dock during visits to change crews, and in recent years the sailing vessels of the Ocean Youth Club have made similar use of the dock. At Timber Quay there is often a Trinity

Pages 78-79: *A scene in the dock near the time of the annual sailing barge match in the Orwell that recalls days when the quays were busy with barges either unloading timber, cement, grain, sand or some other imported cargo or loading malt to be taken to the London breweries.*
Geest North Sea Lines

The mv Harglen *is unloaded by one of the new cranes installed at South-west Quay in 1964.*
Ipswich Port Authority

The Wet Dock seen from the air in October 1969, with the entrance lock at bottom right. The site of the original lock is still apparent towards the left, marked by an indentation in the bank of the New Cut. The mv Arneb *is moored at the ro-ro berth and other vessels are alongside the new Orwell Quay, while a collier is discharging at the gasworks quay.*
Ipswich Port Authority

House lightvessel undergoing conversion to automatic operation, an extension of the activities of the Trinity House depot at Harwich. It was found that the still waters of the dock were preferable to the tidal waters of the moorings in the Stour when intricate electronic work had to be carried out.

On the Island Site there are warehouses, workshops, offices and stores for the Port Authority, parking for some of the many trailers and trucks which bring cargo to and from the port, and premises for several companies performing services in the port area. The large warehouses along the eastern side of the Wet Dock store grain, animal feedstuffs and fertiliser.

Our Victorian forefathers took a brave decision when they resolved to build the largest wet dock in the country. Their foresight laid the foundation for the development of the future Port of Ipswich, which over the past 150 years has brought much prosperity to the town and to the region. We look back with pride and forward with confidence that our successors can continue to build upon this truly great Victorian Vision.

The mv Gulf Sea *taking on cargo for the eastern Mediterranean at Orwell Quay in 1972.*
Ipswich Port Authority

Acknowledgements

It is not to be expected that any book could be written without the help of numerous people, and the authors would wish to thank all who have assisted them with information and advice and with the loan of pictures old and new.

The number of people who have helped makes it impossible to thank each one individually, but the authors would like to express their gratitiude to them all, and particularly to the staffs of Ipswich Port Authority, Suffolk Record Office and Ipswich Museums and Galleries for the interest they have taken in this project as well as for their assistance.

They acknowledge the assistance of David Jones and Hugh Moffat, who both read the manuscript and commented on it, but would point out that any shortcomings or errors are the responsibility of the authors alone.

For providing photographs they thank David Kindred, Richard Smith, the *East Anglian Daily Times*, Geest North Sea Lines, and Ipswich Museums and Galleries and the Suffolk Record Office.

Index

Vessel names are shown in *italics.* Illustrations are listed in **bold** type.